Mark Alpert

Saint Joan of New York

A Novel About God and String Theory

Mark Alpert
New York, NY, USA

ISSN 2197-1188 ISSN 2197-1196 (electronic)
Science and Fiction
ISBN 978-3-030-32552-7 ISBN 978-3-030-32553-4 (eBook)
https://doi.org/10.1007/978-3-030-32553-4

This Springer imprint is published by the registered company Springer Nature Switzerland AG.
The registered company address is: Gewerbestrasse 11, 6330 Cham, Switzerland

Contents

Introduction: Physics and the Search for Meaning

This novel, as the subtitle says, is about God and string theory. Its heroine is a modern-day version of Joan of Arc; she's a New York City teenager who has an extraordinary talent for math and physics. The novel's plot mixes science and religion, which is perhaps a questionable strategy, more likely to upset readers than to please them. When you write about God, you're almost certain to offend someone.

So why did I write such a strange book? It all goes back to the fall of 1981, when I was studying astrophysics at Princeton University.

My faculty adviser at the time was J. Richard Gott III, an expert on Einstein's general theory of relativity. In contrast, I was a neophyte, unskilled but eager. My research project was straightforward: determine how the theory of relativity would work in a hypothetical universe that has only two spatial dimensions (plus the dimension of time). In other words, imagine a cosmic Flatland that resembles a vast sheet of paper, with infinite length and width, but zero thickness. According to Einstein's theory, how would a massive body in this universe affect the space-time around it? Would there be gravity in Flatland?

The math really wasn't that hard. The Einstein Field Equations are much easier to solve in a hypothetical (2 + 1)-dimensional universe than in our actual (3 + 1)-dimensional cosmos. The infamously intricate tensors in those equations—the Einstein tensor (which describes the curvature of space-time) and the stress-energy tensor (which describes the density and flux of energy and momentum)—have only nine components in Flatland, versus the usual sixteen. Even with my poor math skills, I was able to find a solution that showed the shape of (2 + 1)-dimensional space-time around a point mass.

M. Alpert, *Saint Joan of New York*, Science and Fiction,
https://doi.org/10.1007/978-3-030-32553-4_1

Unfortunately, I couldn't figure out the meaning of this mathematical formula. So I showed it to Professor Gott. I rushed to his office and turned to the page in my notebook on which I'd written the space-time metric in pencil. That was the moment when he gave me the best compliment that one theorist can give to another.

"This solution is non-trivial!" he exclaimed.

Basically, we found that there would be no attraction between masses in Flatland, and the geometry of space-time around a point mass would be a cone.[1] More important, though, I experienced the rare, gratifying wonder of making a scientific discovery. The results of our calculations were neither obvious nor insignificant. They revealed something that was both true and surprising, true and beautiful.

In all the years since then, as I've worked as a researcher, science journalist, magazine editor, and novelist, I've been obsessed with non-trivial cosmic truths. How did the universe start, and how will it end? What are the most fundamental laws of physics, and how did they arise? And do those laws show any sign that the universe has a plan, a purpose? I didn't expect to see all those mysteries resolved in my lifetime, but I assumed that scientists would come closer to the answers. Recently, however, physicists have hit a few stumbling blocks. The universe isn't giving up its secrets so easily.

In 1998 I joined the Board of Editors at *Scientific American*, where I oversaw the publication of articles written by some of the world's foremost researchers. That same year, astronomers redrew our picture of the cosmos by discovering that supernovas in distant galaxies were farther away than they'd expected. Their theories had predicted that the expansion of the universe should've slowed down over its 14-billion-year history due to the combined gravity of all its matter, but the supernova results showed just the opposite. The cosmic expansion is actually accelerating.

In an attempt to explain the findings, theorists proposed that an entity called dark energy pervades the universe with a repulsive force that works against gravity. But they could only guess what dark energy is: An inherent property of empty space? A type of dynamic field that's remodeling the universe? And over the past two decades, they haven't come much closer to figuring it out. There's been a similar lack of progress in identifying the nature of dark matter, the unknown substance whose gravity seems to hold galaxies and galactic clusters together. The physicists' ignorance is especially embarrassing

[1] For more details, see "General Relativity in a (2 + 1)-Dimensional Space-Time," J. Richard Gott III and Mark Alpert, *General Relativity and Gravitation*, March 1984, Vol. 16, Issue 3, pp. 243–247 (https://link.springer.com/article/10.1007/BF00762539).

because dark energy seems to constitute 68 percent of the universe's energy content, and dark matter comprises another 27 percent. The total amount of ordinary matter—all the stars, planets, gas clouds, and so on—accounts for only 5 percent. In other words, we understand just a tiny sliver of the cosmos.

Researchers have tried to find answers using the tools of particle physics, but those efforts have fallen short. The Standard Model of particle physics, which is a type of quantum field theory, describes all the known elementary particles and explains three of the four known fundamental forces (electromagnetism and the strong and weak nuclear forces). But it can't explain gravity. Quantum theory, which does an excellent job of describing how particles and forces interact at the atomic and subatomic scales, is mathematically incompatible with general relativity, which explains large-scale phenomena such as galaxies and gravity.

Starting in the 1970s, physicists attempted to merge these two disciplines by developing a theory of quantum gravity that would describe the gravitational force at the smallest scales and highest energies. The most prominent effort involved reimagining the elementary particles as vibrating strings. Dubbed string theory, this project advanced rapidly in the 1980s. The big advantage of string theory is its potential ability to describe *all* particles and forces as different manifestations of a fundamental one-dimensional object, a string that is infinitely slender and only 10^{-35} meter long (that is, about a trillionth of a trillionth of a trillionth of a meter). A closed string—that is, a minuscule loop—would have the properties of a graviton, the hypothesized carrier of the gravitational force. An open string, with two endpoints, could describe any of the other particles, depending on its mode of vibration; a string vibrating in one type of pattern would be an electron, for example, and strings vibrating in other patterns would be quarks, photons, neutrinos, and so on.

String theory, though, has big disadvantages too. To generate all those particles and forces, the infinitesimal strings must vibrate in nine spatial dimensions, six more than the number we've observed in our (3 + 1)-dimensional universe. String theorists deal with this discrepancy by postulating that the extra dimensions are curled up into manifolds too microscopic to be observed, like the tiny curls of fabric in a seemingly flat carpet. But when physicists tried to derive equations and predictions using this approach, it proved to be hideously complex. Instead of discovering a unique Theory of Everything, they developed five intriguing but incomplete theories. In each theory, moreover, there's a huge number of ways to fold up the extra dimensions. And because the shape of each possible manifold would determine the properties of all the strings vibrating within it, the theory allows for a vast array of possible universes instead of specifically describing the universe we inhabit.

In the 1990s researchers proposed that all the ten-dimensional string theories were part of an eleven-dimensional framework called M-theory. (The M stands for mystery, magic, or membrane, depending on whom you ask.) But M-theory is maddeningly sketchy. So far, it hasn't yielded any equations describing our universe, much less any predictions. Physicists have focused instead on studying the "string landscape," the collection of all the universes that can be created by varying the theory's parameters.

This landscape is immense. Estimates of the number of theoretical possibilities range as high as $10^{272,000}$. What's more, each possible universe in the landscape could be just as real as our own. According to the theory of inflation, which cosmologists developed to explain the origins of the Big Bang, a swiftly expanding proto-universe not only gave rise to our cosmos, but it might be perpetually generating other universes as well, like bubbles in boiling water. Each universe in this so-called multiverse might have radically different physical laws and constants, perhaps dictated by its position in the string landscape. The wild expansion of the multiverse would keep the bubble universes well separated, preventing us from ever visiting any of the other realms, but our bubble could've collided with another while they were forming, leaving an imprint on our cosmos that we might be able to detect.

Some scientists welcomed the multiverse concept because it would explain why so many of the physical constants in our universe seem to be fine-tuned—that is, their values are within the ranges that allow for the creation of stars, planets, and life. If, for example, the fine-structure constant (which specifies the strength of the electromagnetic force) were much larger, atoms couldn't form; if it were much smaller, stars couldn't shine. This fine-tuning unsettles many physicists because it seems to imply that either the universe was designed to support life, or we are the lucky beneficiaries of an unlikely cosmic accident. But if a multiverse truly exists, and if each universe in the vast landscape has a different set of physical constants, then it would be no surprise to find ourselves in one of the few universes suited for life, because no observers would ever evolve in the less hospitable universes.

Other physicists, however, believe the multiverse idea is unscientific, because it probably can't be tested. If no other universes ever collided with ours, how can we prove they exist? Furthermore, some researchers have recently proposed that string theory can't predict a universe like ours—a cosmos expanding at an accelerating rate—because those kinds of theoretical possibilities are logically inconsistent. (This proposal has been called the swampland conjecture, since it posits that our universe lies in a mucky, nonviable part of the string landscape.) And other physicists are reluctant to give up their long-held dreams of a unique Theory of Everything. They still yearn to explain gravity, quantum fields, and all the physical constants as the inevitable consequences of a beautiful, unified theory that can be expressed as a set of solvable, testable equations.

Could string theory or M-theory ultimately lead us to this putative Theory of Everything? String theorists had expected to find support for their approach in the recent experiments conducted at the Large Hadron Collider (LHC), the world's most powerful particle accelerator, a 27-kilometer underground ring that straddles the border between France and Switzerland. Starting in 2009, researchers at the LHC accelerated beams of protons around the giant ring and smashed them together at velocities very close to the speed of light, producing collisions with energies as high as 13 trillion electron volts (TeV). The debris from the high-energy proton impacts can reveal the existence of new particles; in 2012, for example, LHC researchers detected the Higgs boson, which was the last particle predicted by the Standard Model to be discovered.

But the great hope of the string theorists was that the LHC would also reveal supersymmetric particles. String theory is built upon the principle of supersymmetry, which stipulates that every particle described in the Standard Model must have a heavier partner. If string theory is indeed a correct description of the universe, those superpartner particles must exist. So far, though, LHC researchers have found no evidence of the superpartners in the collision debris. This negative result doesn't necessarily mean that the supersymmetric particles don't exist; they may be so massive that the proton collisions in the LHC aren't energetic enough to produce them. The findings, however, have ruled out the simplest and most elegant versions of supersymmetry, and the prospects of string theory seem dimmer as a result.

What makes physicists even more worried is that they're not getting enough experimental data to guide the development of their theories. Conducting experiments that gather new information about fundamental physics—discoveries about particles, forces, cosmology, and space-time—is growing more difficult and expensive. It cost $8 billion to build the LHC, and constructing a next-generation collider that could search for particles at higher energies could cost three times as much. In a way, physics is a victim of its own success: all the easy experiments have already been done. Gaining new knowledge will require bigger tools and cleverer studies.

And even if the scientific community somehow finds the money to build a 100-kilometer accelerator ring that could produce proton collisions with energies up to 100 TeV, there's no guarantee that those experiments will reveal new phenomena. The supersymmetric particles, if they exist, might well be more massive than 100 TeV, in which case the next-generation collider would fail to discover them. To thoroughly plumb the details of fundamental physics, researchers would have to produce collisions that approach the Planck energy, which would bend space-time violently enough to reveal the effects of quantum gravity. The Planck energy, however, is about *a quadrillion times higher* than the LHC's collision energies. Reaching that threshold would require a particle collider the size of the Milky Way galaxy.

Fortunately, there's another way to catch a glimpse of Planckian physics. In the first moments of the Big Bang, matter and energy were so compressed that quantum gravity might have influenced the early history of the universe. In particular, it might've left an imprint on the cosmic microwave background (CMB), the radiation emitted when the first hydrogen atoms formed and the universe turned transparent, which occurred 380,000 years after the Big Bang started. New telescopes are trying to detect a special kind of polarization of the CMB—called B-mode—which would indicate the presence of gravitational waves produced by the hypothesized process of inflation. Earlier searches for this CMB polarization (notably, the BICEP2 results reported in 2014) were marred by stray microwaves reradiated by galactic dust, but the newer observatories might detect the subtle cosmic signals hiding amidst the noise. If not, perhaps the primordial gravitational waves could be revealed by the Laser Interferometer Space Antenna, a proposed space-based detector scheduled to be launched in the 2030s.

Other clues to fundamental physics might come from ongoing studies of dark energy, the mysterious entity that seems to be speeding up the expansion of the universe. Astronomers are carefully measuring the spatial distribution of galaxies and galactic clusters to determine how the rate of cosmic expansion has varied over the past 14 billion years. More precise measurements of distances and redshifts will come from planned space telescopes such as Euclid and WFIRST. Investigations of dark energy and dark matter might indicate that gravity works differently at galactic distances than it does at smaller scales. If that's true, scientists will need to revise the general theory of relativity.

Until those experimental results come in, however, physicists may have to endure a long period of uncertainty and stasis. Although string theorists continue to propose new ideas, and other researchers are working on alternative approaches—loop quantum gravity, asymptotically safe gravity, causal dynamical triangulations, and so on—the search for a fundamental theory will be difficult without guidance from new data. How can scientists decide if a theory is worth pursuing if they don't have enough facts to prove it right or wrong? And here's the nightmare scenario for fundamental physics: what if certain crucial facts are simply unattainable?

In the absence of experimental data, some theorists appear to be using dubious criteria when choosing which theories to work on. Sabine Hossenfelder, a research fellow at the Frankfurt Institute for Advanced Studies, notes that many physicists have a bias against theories that seem to have unlikely numerical coincidences. For example, some theorists believe the Standard Model is "ugly" because it requires the near-cancellation of two very large and almost equal parameters to calculate the mass of the Higgs boson. But in her recent book, *Lost in Math: How Beauty Leads Physics Astray*, Hossenfelder points out that you can't say the Standard Model's parameters are unlikely if you don't know their probability distribution. She argues that the general preference for "natural" theories that have no awkward-looking terms is really an aesthetic choice. Physicists want to work on beautiful theories, but beautiful ideas aren't always correct.

Beauty is in the eye of the beholder, of course, and our appreciation of it is linked to our sense of wonder. So I would go a step further than Hossenfelder and declare that a more primitive instinct is motivating physicists to construct elaborate mathematical frameworks that may have little or no connection to the real world. Although most scientists would hotly dispute this characterization, they seem to be searching for divine order in a messy universe.

* * *

I should make it clear at this point that I have no religious agenda. I'm not a believer. I'm not a committed atheist either. When I worked as a researcher and a science journalist, my job was ferreting out the truth, no matter where it led. *Scientific American* was especially diligent about exposing the falsehoods of "Intelligent Design" proponents who claimed to see God's hand in the fashioning of complex biological structures such as the human eye and the bacterial flagellum. In 2002 we published "15 Answers to Creationist Nonsense," which demolished the unscientific arguments against evolution and became one of the most widely shared articles in the magazine's history.

But after ten years as an editor at *Scientific American*, I stepped away from journalism and started writing fiction. I wrote novels about physics and Albert Einstein and quantum theory. And though I'm not a big fan of organized religion (and my position on the moral spectrum is definitely below average), ideas about God keep popping up in my books.

I've been inspired by Flannery O'Connor, the Georgia-born, mid-twentieth-century author famous for her Southern Gothic fiction and her fixation on religious themes. In her short story, "A Good Man is Hard to Find," O'Connor focuses on the Misfit, a murderer who escapes from prison and encounters an ordinary (but very dysfunctional) family of tourists on a lonely country road. While the Misfit's henchmen escort the family members into the woods, one by one, to be executed, the family's grandmother pleads for her life by appealing to the criminal's belief in Jesus. "Pray!" she cries. "Jesus, you ought not to shoot a lady." The Misfit, though, is unmoved. He seems troubled by the possibility that God exists, but he says he wouldn't change his ways unless he personally witnessed Jesus raise the dead. "I wisht I had of been there," he says. "If I had of been there I would of known and I wouldn't be like I am now." Then he shoots the grandmother three times in the chest and starts to clean his glasses.

Like the Misfit, scientists yearn for hard facts about mysterious phenomena. But should they even try to answer questions about the purpose of the universe? Most researchers operate under the assumption that science and religion are completely separate fields—or, in the phrase coined by evolutionary biologist Stephen Jay Gould, "non-overlapping magisteria." According to this premise, the tools of science can't answer questions of faith, and religious beliefs shouldn't influence the scientific method. But as physicists investigate the most fundamental characteristics of nature, they're tackling issues that have long been the province of philosophers and theologians: Is the universe infinite and eternal? Why does it seem to follow mathematical laws, and are those laws inevitable? And, perhaps most important, why does the universe exist? Why is there *something* instead of *nothing*?

These questions are similar to the ones that medieval philosopher St. Thomas Aquinas tried to resolve in the thirteenth century. In his book *Summa Theologica*, Aquinas presented five arguments for God's existence, which he called the Five Ways. In his first argument, he observed that all worldly objects can change from potential to actuality—an ice cube can melt, a child can grow—but the cause of that change must be something besides that object (warm air melts the ice cube, food nourishes the child). The history of the universe can thus be seen as an endless chain of changes, but Aquinas argued that there must be some transcendent entity that initiated the chain, something that is itself unchanging and already possesses all the properties that worldly objects can come to possess. In his second argument, he claimed that this entity must be eternal; because it is the root of all causes, nothing else could've caused it. And in the third argument, Aquinas added that unlike all worldly objects, which may or may not come into existence, the transcendent entity is necessary—it *must* exist.

Aquinas defined that entity as God. Over the following centuries this line of reasoning came to be known as the cosmological argument, and many philosophers elaborated on it. In the late seventeenth century, German philosopher Gottfried Leibniz was the first to explicitly ask, "Why is there something instead of nothing?" To answer the question, he proposed the principle of sufficient reason, which states that there must be a reasonable explanation for every entity and phenomenon in the universe, but the ultimate and sufficient reason for all worldly things must lie outside the long succession of contingent causes and events. According to Leibniz, this ultimate reason is God, whom he described as "the necessary being which has in itself the reason for its existence." It's interesting to note that Leibniz was also a mathematician and physicist; in fact, he invented differential and integral calculus at about the same time that Isaac Newton did. (They developed the math independently.) Both Leibniz and Newton considered themselves natural philosophers, and they freely jumped back and forth between science and theology.

By the twentieth century, most scientists no longer devised proofs of God's existence, but the connection between physics and faith hadn't been entirely severed. Einstein, who frequently wrote and spoke about religion, didn't believe in a personal God who influences history or human behavior, but he wasn't an atheist either. (I'm focusing on Einstein because I wrote a novel about him. I've internalized his point of view.[2]) He preferred to call himself agnostic, although he sometimes leaned toward the pantheism of Jewish-Dutch philosopher Baruch Spinoza, who proclaimed in the seventeenth century that God is identical with nature. Likewise, Einstein compared the human race to a small child in a library full of books written in unfamiliar languages: "The child notes a definite plan in the arrangement of the books, a mysterious order, which it does not comprehend, but only dimly suspects. That, it seems to me, is the attitude of the human mind, even the greatest and most cultured, toward God. We see a universe marvelously arranged, obeying certain laws, but we understand the laws only dimly."

Einstein often invoked God when he talked about physics. In 1919, after British scientists confirmed Einstein's general theory of relativity by detecting the bending of starlight around the sun, he was asked how he would've reacted if the researchers hadn't found the supporting evidence. "Then I would have felt sorry for the dear Lord," Einstein said. "The theory is correct." His attitude was a strange mix of humility and arrogance. He was clearly awed by the

[2] *Final Theory*, Mark Alpert (Simon & Schuster, 2008).

laws of physics and grateful that they were mathematically decipherable. ("The eternal mystery of the world is its comprehensibility... The fact that it is comprehensible is a miracle.") But during the 1930s he fiercely opposed the emerging field of quantum mechanics because it clashed with his firm beliefs about the universe.

Einstein assumed that the world is deterministic—that is, physical actions always have predictable effects. Quantum theory, though, doesn't make exact predictions about particle interactions. The theory describes every particle as a wave function, which doesn't specify the quantum state of the particle; instead it provides a distribution of probabilities for the particle's position, momentum, and other observable properties. So theorists, for example, can predict the chance that an interaction such as radioactive decay will happen within a certain amount of time, but they can't predict exactly when a particular atom will decay. Einstein famously criticized the indeterminacy of quantum theory by saying, "God does not play dice with the universe." (Niels Bohr, the father of quantum mechanics, famously replied, "Einstein, stop telling God what to do.")

Although quantum theory is now the foundation of particle physics, many scientists still share Einstein's discomfort with its implications. According to the wave-function formulas, a particle can be in more than one quantum state at the same time—its spin, for example, can be oriented upward and downward simultaneously. But this so-called superposition only lasts until an observer measures the particle's properties. In that instant, the wave function "collapses" to a single state; the particle's spin, for example, becomes oriented either upward or downward. (Measurements can also clarify the particle's position and momentum, but only within the limits of the uncertainty principle, which states that both properties can't be precisely known at the same time.) The theory thus incorporates observers into its framework, because they play an important role in quantum processes. But how would the theory work in a universe without observers? And how exactly does the act of observation change the physical state of a particle?

Bohr and other physicists dodged these questions by arguing that a more complete, deterministic description of particles might not be possible. In their view, which is often called the Copenhagen interpretation, quantum theory explains our observations of the subatomic world without detailing all its inner workings, and this explanation might be the best we can hope for. And because the quantum formulas calculate the probabilities of particle interactions so well, many physicists urged their skeptical colleagues to simply "shut up and calculate." But other researchers sought different explanations, hoping to shore up the logical foundations of quantum theory or perhaps discover a more fundamental framework beneath it. Hugh Everett, for example, proposed the many-worlds interpretation, which claims that the wave function doesn't collapse at the moment of observation; instead, each possible outcome of a particle measurement occurs in an alternate reality. In our world, we detect that the particle's spin is pointing up, but at that moment another reality splits off from ours, and in that world our alternate selves detect that the particle's spin is pointing down.

And there are many more examples of quantum weirdness. Physicists can put two particles in a state of quantum entanglement, which forces the properties of one particle to correlate with those of the other. In an entangled pair of electrons, for instance, if one particle is in a superposition of two spin states (say, up/down), then the other particle must be in the complementary superposition (down/up). This correlation will persist as long as neither electron interacts with its environment, even if the researchers separate the entangled particles by thousands of kilometers. But if an observer measures the spin of one particle and its wave function collapses to the up-spin state, then the wave function of the other particle will collapse to the down-spin state *instantaneously*. This violates the principle of locality—how can the effects of a physical event travel across the universe faster than the speed of light? Somehow, quantum entanglement can create a direct connection between far-flung parts of the cosmos. Einstein derisively called it "spooky action at a distance," but many experiments have demonstrated the phenomenon.

In short, physicists have revealed aspects of nature that seem supernatural. We live in a universe where the act of observing something can alter its reality, and where distant pieces of space-time can be woven together. Its laws are eerily mathematical, but when we try to apply them to certain situations (small scales, high energies) the equations sometimes yield nonsensical infinities. Worse, those laws put some strict limits on what we can learn about the universe. We probably can't test the multiverse hypothesis or the many-worlds interpretation, because those parallel realms are unlikely to be observable. And here in our own corner of the cosmos, we can't peer inside black holes or view anything that lies beyond the distance that light has traveled since the start of the Big Bang. Despite centuries of scientific exploration, we still face confounding mysteries. I can't help but think of the traditional fisherman's prayer: "Oh God, thy sea is so great and my boat is so small!"

Is there a place in this universe for the causative God of Aquinas and Leibniz? Or maybe the more diffuse and impersonal God of Spinoza? Particle physicist Victor Stenger addressed this question in his 2007 book *God: The Failed Hypothesis*. (To make his position absolutely clear, he gave the book the subtitle "How Science Shows That God Does Not Exist.") Stenger quickly dismisses the theist notion of a God who responds to prayers and cures ill children, because scientists would've noticed that kind of divine intervention by now. Then he argues, less convincingly, against the existence of a deist God who created the universe and its laws and then stood back and watched it run.

Stenger claims that the fundamental parameters of the universe that determine the strengths of the forces and the masses of the particles aren't so fine-tuned after all. The physical constants, he says, could've been set at substantially different values and still allowed for a cosmos capable of supporting life. What's more, he contends that many laws of nature (such as the conservation of energy) follow inevitably from the observed symmetries of the universe (there's no special point or direction in space, for example). Other laws resulted from the spontaneous breaking of symmetries that prevailed during the explosive start of the Big Bang. Stenger concludes: "There is no reason why the laws of physics cannot have come from within the universe itself."

Explaining the creation of the universe is trickier, though. Cosmologists don't know if the universe even had a beginning; instead, it might've had an eternal past *before* the Big Bang, stretching infinitely backward in time. Some cosmological models propose that the universe has gone through endless cycles of expansion and contraction, and some versions of the theory of inflation postulate an eternal process in which new universes are forever branching off from the speedily expanding "inflationary background." But other cosmologists argue that inflation had to start somewhere, and the starting point could've been literally *nothing*. As we've learned from quantum theory, empty space isn't totally empty; the vacuum has energy—a small amount, but detectable—because it's always churning with virtual particles briefly popping in and out of existence. In other words, nothingness is unstable. Over a long stretch of time, all kinds of improbable things can happen in empty space, and one of them might've been a sudden drop to a lower vacuum energy, which would've triggered the exponential expansion. In the most mind-blowing example of quantum weirdness, a random fluctuation in a speck of primordial emptiness could've set off the furious growth of the multiverse.

For Stenger, this theoretical possibility is evidence that God isn't needed for Creation. "The natural state of affairs is something rather than nothing," he writes. "An empty universe requires supernatural intervention—not a full one." But this conclusion seems a bit hasty. Scientists don't fully comprehend the quantum world yet, and their hypotheses about the first moments of Creation aren't much more than guesses at this point. We need to discover and understand the fundamental laws of physics before we can say they're inevitable. And we need to explore the universe and its history a little more thoroughly before we can make such definitive statements about its origins.

Just for the sake of argument, though, let's assume that this hypothesis of Quantum Creation is correct. Suppose we do live in a universe that generated its own laws and called itself into being. Doesn't that sound a lot like Leibniz's description of God ("the necessary being which has in itself the reason for its existence")? It's also similar to Spinoza's pantheism, his proposition that the universe as a whole is God. Instead of proving that God doesn't exist, maybe science will broaden our definition of divinity. The Universal God might provide less solace than the traditional version does; there's no love or goodness in the new definition, no place for immortal souls or a metaphysical afterlife. Pantheism, however, might offer its own benefits. If we treated every part of our planet as a manifestation of God—every person, animal, forest, and river—then perhaps we wouldn't be so quick to ravage our environment and our fellow humans.

But let's not get ahead of ourselves. If we truly want to contribute to humanity's search for meaning, we should prioritize the funding of advanced telescopes, detectors and other scientific instruments that can provide the desperately needed empirical data about particles, forces, space-time, and the history of the universe. Once researchers have more facts to analyze, they're bound to find more clues to fundamental physics. And until the new observations come in, I believe the priority for theorists should be to rethink the logic behind quantum theory. Are its assumptions sound? Are alternative formulations possible? Reexamining these foundational questions might help resolve some of the troubles confronting string theory, which is based on quantum principles.

Maybe this effort will lead to breakthroughs in theology as well. The pivotal role of observers in quantum theory is very curious. Is it possible that the human race has a cosmic purpose after all? Did the universe blossom into an untold number of realities, each containing billions of galaxies and vast oceans of emptiness between them, just to produce a few scattered communities of observers? Is the ultimate goal of the universe to observe its own splendor?

Perhaps. We'll have to wait and see.

* * *

In conclusion, it seems clear that the theologians haven't proved the existence of God, but the physicists haven't disproved it either. In a 1952 essay titled "Is There a God?" British philosopher Bertrand Russell argued that the burden of proof lies with the believers rather than the skeptics. Russell compared the God hypothesis with another claim that can't be proved false: that a china teapot might be floating in outer space, orbiting the sun between Earth and Mars. Because this hypothetical teapot would be too small to be spotted by any telescope, astronomers could never rule out its existence, and yet any reasonable person would dismiss the possibility. ("A teapot in space? How ridiculous! How would it get there?") Other philosophers have noted, however, that the analogy between God and the orbiting teapot is flawed. If God is a transcendent entity that lies beyond the realm of empirical detection, then faith in such a deity is very different from a belief in celestial chinaware.[3]

So, after hundreds of years of debate and scientific discoveries, there's still plenty of room for choice when it comes to religious beliefs. And that's the point I wanted to make when I wrote my novel *Saint Joan of New York*.

I used the Joan of Arc story as the framework for this book because it nicely dramatizes the conflict between faith and doubt. In the small French village of Domrémy in the early fifteenth century, a teenage peasant girl named Jeanne d'Arc told her family and friends that she'd had spiritual visitations from three key figures of medieval Christianity: Saint Margaret, Saint Catherine, and the archangel Michael. They'd spoken with her many times, she said, usually while she tended her family's flock of sheep. And they'd given Jeanne a holy mission: God had commanded her to anoint the French king and help him drive the English invaders out of the country. (France and England were fighting the Hundred Years' War, which was going very badly for France. The previous French king had gone mad and lost much of his kingdom to the English. His son, Charles VII—called the Dauphin because he hadn't been crowned yet—had retreated to central France and considered fleeing the country altogether.)

Now remember, this was the Middle Ages, and reports of divine visitation were taken much more seriously in those days. Nevertheless, most of Jeanne's neighbors in Domrémy must've thought she was insane.

But by 1428 she'd convinced her uncle to take her to the nearby town of Vaucouleurs to see Robert de Baudricourt, a military captain who could arrange safe passage to the French royal court. At first Baudricourt dismissed Jeanne and sent her back to Domrémy, but she kept returning to entreat him. In 1429 he finally relented, ordering his knights to escort her to the Dauphin. At the French court she won over Charles VII, who allowed her to join his army and fight the English soldiers besieging the city of Orléans. Inspired by Jeanne's piety and bravery in battle, the French troops lifted the siege and then won victory after victory against the English. Under her direction, the French

[3] *Is God a Delusion?*, Eric Reitan (Wiley-Blackwell, 2008).

army marched in triumph to the cathedral city of Reims, where Charles VII was anointed with oil and crowned.

The obvious question is, how did she do it? In a matter of weeks an illiterate seventeen-year-old with no military training took control of an army. She mastered the skills of leadership, horsemanship, and medieval combat. (According to her fellow commanders, she was particularly expert at arranging the placement of cannons, which were relatively new to European warfare.) She clearly had some extraordinary qualities—intelligence, charisma, fortitude—but are they enough to explain her astounding success? And in 1431, after her enemies captured her in battle and put her on trial for heresy, where did she find the strength and cleverness to endure the many weeks of interrogation, not to mention her excruciating execution? If God wasn't guiding her, what was?

Of course, I'm not the first writer to reexamine this story. Mark Twain spent fourteen years researching and writing his novel *Personal Recollections of Joan of Arc*. George Bernard Shaw's play *Saint Joan* premiered in 1923, three years after the Catholic Church finally canonized her. (Joan's sainthood was delayed for several centuries, partly because the church worried about antagonizing the English.) Twain and Shaw had a wonderful resource to help them illuminate Joan's character: the preserved transcripts of the 1431 trial that sentenced her to death and the retrial that posthumously declared her innocent 25 years later. Thanks to those records, historians and writers know Joan better than they know almost anyone else who lived during the Middle Ages.

But I think I can safely say that I'm the first writer to reimagine Joan as a math prodigy. Instead of portraying a conflict between nations, I wanted to show a war of ideas, and fundamental physics has become a battleground in that war. Scientists are exploring the foundations of reality, and believers and atheists are analyzing the discoveries, looking for anything that can buttress their philosophical arguments. I saw Joan as a mediator in that conflict, someone who could bring the two sides together. As a result, there's a good chance that neither the religious nor the nonreligious will be satisfied with the story I've told. But that's okay. You can read it any way you want to.

The best argument for God's existence, I think, isn't in the Bible or any theological tract. It isn't in any work of science or philosophy either. It's in Yann Martel's novel *Life of Pi*. This book tells the improbable story of Pi Patel, a religious young man who survives a shipwreck and floats across the Pacific Ocean in a lifeboat he shares with a Bengal tiger. When he finally reaches land, the tiger disappears into the jungle, and when Pi is questioned about the shipwreck, he explains how he managed to live in close quarters with a man-eating predator for seven months. His interrogators are understandably skeptical of his tale, so after some prodding he gives them an alternative story of the shipwreck. This account is more gruesome than the first story, but also more believable, because it omits the tiger. Seeing that his questioners are now satisfied, Pi asks them which is the better story, the one with the tiger or the one without it. The interrogators admit that the first story is better. Pi responds, "And so it goes with God."

What about the Joan of Arc story? Is it more interesting to imagine that God inspired and sustained her, or that Joan herself dreamed up a glorious plan and pursued it with every ounce of her will, changing history in the process? Does God make the story better or worse?

I'll leave it for you to decide.

Saint Joan of New York
A Novel About God and String Theory

Part One: The Magic Number

Chapter One

I'll start at the beginning, okay? My name is Joan Cooper, and I'm seventeen years old. I'm going to tell you about the first time I saw God.

It happened last October, on a Saturday afternoon. I was in the Bronx, running in the biggest race of the cross-country track season, the city championship for New York's high schools. And I was doing great, even better than I'd hoped. I sprinted down the trail through the woods of Van Cortlandt Park, way ahead of all the other girls on the five-kilometer course. By the time I reached the halfway point, I had a fifty-yard lead on the pack behind me.

Seriously, I was killing it. I felt strong, pumped. But most of all, I felt relieved. The beginning of my senior year had been a nightmare. I hadn't smiled in months. But now I was practically laughing as I charged up and down the wooded hills. I knew I was going to win the race, and winning still felt good.

That's when I saw the Lord Almighty, although I didn't realize at the time that I was looking at the Creator. I thought I saw a fallen runner, a puny African-American boy who'd collapsed on the trail.

He was at the bottom of the course's steepest hill, a hundred feet ahead. The boy lay facedown on the edge of the trail, half-on and half-off the path, his skinny legs splayed across the mud and dead leaves. He wore a team uniform—bright red shorts and a sleeveless track shirt—so I assumed he was one of the runners in the Boys Junior Varsity race, which had started twenty

M. Alpert, *Saint Joan of New York*, Science and Fiction,
https://doi.org/10.1007/978-3-030-32553-4_2

minutes before the Girls Varsity. I figured he must've been in last place, lagging behind all the other JV boys, and then he'd slipped on the steep, muddy slope, and nobody had seen him fall. And because I was ahead of everyone else in the next race, I was the first person to run into him.

At first I was just startled. But as I got closer and took a better look at the kid, I noticed he wasn't moving. It was a frigid day for October, barely forty degrees, and yet the boy hadn't curled up into a ball for warmth. His head was bent at a sharp angle, skewed against his left shoulder.

I started to panic. Because I thought of my sister.

I hurtled downhill and stopped next to the kid. My head swam as I looked down at him, and my legs trembled. I tried to calm myself by doing some fast calculations in my head—Newton's Second Law, gravity and acceleration, the maximum force of the impact when the boy hit the ground. I'm a math geek, a nerd to the core, and when I get nervous I can usually calm myself down by crunching numbers, calculating probabilities. But this was bad, *really* bad. I couldn't stop shaking.

I crouched beside him. "Hey? You okay?"

He didn't answer. The boy lay still, his shorts and shirt splattered with mud, the back of his head covered with the black stubble of a buzz cut. He was tiny, smaller than any JV runner I'd ever seen, more like a middle-school kid than a high-schooler. I was afraid to touch him—it looked like his neck might be broken—but I put my hand on his bare shoulder and gave it a light squeeze. "Hey, kid? Can you hear me?"

Still no answer. I heard a clattering behind my back, the pounding footsteps of the pack of runners catching up to me, but when I looked up I saw that no one else was stopping. Seven of the fastest Varsity girls from Brooklyn and Manhattan zipped right past me and dashed up the next hill on the course, their eyes fixed straight ahead. Even Elena, my friend and teammate from Franklin High—the Rosalind Franklin High School for Math and Science—refused to break stride.

I couldn't believe it. *The race didn't matter anymore!* Someone needed to stay with this kid while I ran for help. The closest race officials were at the two-kilometer checkpoint, which was several hundred yards behind us.

I stood up and cupped my hands around my mouth, ready to order the girls to come back. But before I could shout anything, the boy rolled over and smiled.

"Don't bother. I'm all right."

His voice was high-pitched, like a little kid's, but strong and cheerful. He raised his head and propped himself up on his elbows, smiling as if he recognized me.

It took me a couple of seconds to get over my surprise. I gave him a once-over, glancing at his head and legs and arms. No cuts, no bruises. Like everyone else in the race, he had a square piece of paper with his race number—137—safety-pinned to the front of his track shirt. There was a big splotch of mud on his shirt that made it hard to read the name of his high school, but his face was spotless, as smooth and pretty as a doll's. His lips were shiny and delicately curved, as if they'd been painted on his mouth. His eyelashes were long and fluttery.

I stared at him. He was angelic. There was no other word for it.

I leaned over him, bending low. "Listen, are you hurt? Did you twist your ankle?"

He shook his head. He seemed amused by my concern. "I'm fine. Perfect, in fact." The boy lifted one of his skinny legs and flexed his ankle, raising and lowering the toe of his muddy sneaker. "See? Nothing broken."

But my heart was still pounding. If the kid wasn't hurt, why was he lying on the edge of the trail? And smiling at me like that? Was this some kind of joke?

"Well, if nothing's wrong, can you get on your feet?" I stretched my hand toward him. "Come on, I'll give you a—"

"No, I'm good. I like the view from down here." Still smiling, he pointed straight up. "The trees are beautiful, right? All the leaves falling. So many colors."

I frowned. This kid was messing with me. He'd already ruined my race—there was no way I could catch up to the pack of girls in the lead—and now he was playing some stupid game, making me feel like an idiot for stopping to help him. I heard more footsteps behind me, and then the second wave of Varsity runners came running down the hill, a tight pack of nine girls. In a few seconds they rushed past, not even glancing at me, just like the first pack. As if I weren't there.

It was odd. And disturbing.

The boy stopped smiling and looked me in the eye. "Stand up, child. You've passed the test. I am well pleased with you."

His voice was serious now, deep and loud. It sounded like someone else had started speaking through the kid, someone much older and bigger, an invisible giant who was using the boy like a ventriloquist's dummy. His words echoed against the hills. The falling leaves seemed to linger in mid-descent, the curled scraps of yellow and orange drifting ever so slowly to the ground.

The boy cocked his head, gesturing toward the trail and the runners in the distance. "Go on, finish your race. But we'll talk again soon. The end of all things is at hand."

I stood up straight, my legs trembling again. All of a sudden, I was terrified. My throat tightened and my mouth went dry. "Okay…okay…I'm going."

I backed away from him. After a while I started jogging down the trail, slowly picking up speed, but I kept looking over my shoulder, swiveling my head every few seconds. The kid had freaked me out. I didn't want to turn my back on him.

I still didn't realize he was God. To tell you the truth, the idea hadn't even occurred to me yet. I felt no solemn awe in his presence, no sense of "divine majesty" or anything hokey like that. But you know how everyone in church says that true believers are supposed to fear the Lord? That's something I *did* feel. I was afraid of that boy lying on the ground. And I was even more afraid that I was going crazy.

I kept my eyes on the kid until I reached the top of the next hill and couldn't see him anymore. Then I faced forward and ran as fast as I could.

Chapter Two

The rest of the race was a blur. I couldn't focus. I passed a few runners on the last hills of the course, and maybe I could've passed a few more if I'd really tried, but it seemed so pointless. I had no enthusiasm for it anymore. I was too confused.

My head ached as I ran the last mile. I couldn't stop thinking about the boy. I saw him again in my mind's eye, lying facedown on the dead leaves. Then I saw him rolling over and turning his face toward me, that beautiful, angelic face.

But there was something terrible about it too. His face reminded me of my sister's.

Samantha was two years older than me. She was also a much better person—kinder, more popular, and a whole lot prettier. I'm just a skinny twerp with ratty black hair, but Samantha was tall and blond and had a terrific body. She got accepted into Princeton, and the college actually put a photo of her on its website, a picture of her running across a soccer field, her blond ponytail swinging behind her. She was so gorgeous, it should've made me resentful, but I loved her too much to be jealous. She'd been my hero since preschool. When I was deciding last spring which colleges to apply to, I put Princeton at the top of my list. I still wanted to be near her.

On June 29, though, her photo disappeared from Princeton's website. They took it down the day after the accident. The same picture of Samantha had already run on the front page of the *New York Post.*

I shook my head as I ran, trying to forget that photo. I stared at the trail instead, my sneakers splashing in the mud. I needed to think about something else, *anything* else.

So I thought about numbers. For a math geek like me, that was a good way to relax. The race number pinned to my track shirt was 418. Not a prime number, obviously, and not a square or a cube either. But if you look at any number closely enough, you can usually find something interesting about it. Divide 418 by two, and you get 209. Which is the product of two prime numbers, eleven and nineteen.

I thought it over for a moment, then tried adding all the prime factors together. Two plus eleven plus nineteen equals thirty-two. After another moment I realized that the product of the digits in 418—four times one times eight—is also equal to thirty-two. So now I'd proved it: 418 *was* an interesting number.

But that was too easy. I was still a long way from the finish line, and I needed a more difficult problem to distract me. I thought of other numbers—my student ID, the address of my family's apartment building, the latitude and longitude of Van Cortlandt Park. I started mentally reciting the digits of π, which I'd memorized to a hundred decimal places. The numbers swirled inside my head as I dashed out of the woods and into the broad, grassy field at the end of the racecourse.

I squinted at the finish line, two hundred yards ahead. The half-dozen Varsity frontrunners had already crossed the line and were staggering toward the cool-down area just beyond it. My teammate Elena finished a few yards behind them, and a couple of girls from Brooklyn Latin High School were a hundred-fifty yards farther back, battling for eighth place. The final stretch was a wide gravel path, and on both sides of it were crowds of spectators, hundreds of screaming teenagers and their coaches. I usually loved this part of the race, the cheers roaring in all directions and propelling me toward the finish, but at that moment I hardly noticed the noise. With sudden alarm, my mind had focused on one particular number, which I'd seen just a few minutes ago: 137, the race number of the boy on the trail.

The more I thought about the number, the more worried I got. It's prime, of course, the first in a pair of twin primes (137 and 139). But it's a very special kind of prime number, a Pythagorean prime, because it's equal to the sum of two squares, 16 and 121. It's also a Stern prime, an Eisenstein prime, and a strictly non-palindromic number. The weirdest thing about it, though, is its connection to the laws of nature. The reciprocal of 137—that is, 1/137—is very close to the fine-structure constant, a quantity that appears in the most fundamental equations of quantum physics. It's a magic number, a key to understanding the universe.

I told myself to calm down. It was just a coincidence. The weird boy had a weird number pinned to his shirt. So what?

But my stomach clenched. I ran headlong down the final stretch, remembering the terror I'd felt in the woods.

In the last twenty yards I passed both of the girls from Brooklyn Latin. I sprinted like crazy toward the finish line, but I didn't look up at the digital clock to see how fast I'd run, because I didn't care about my race time or what place I came in. I kept running at full speed even after I crossed the line, barreling past the cool-down area and the race officials in their fluorescent-yellow vests. They waved their arms at me and screamed, "Slow down!"

Frantic, I looked over my shoulder. I saw the Brooklyn Latin girls finish the race and double over, gasping. I saw the Varsity frontrunners celebrating on the sidelines, the first-place finisher from Beacon High School hugging the girls from Millennium High and Stuyvesant. Everything was fine, normal, just like every other cross-country race I'd ever seen. But I kept running. I was too scared to stop.

I didn't slow down until I reached the corner of the park that was unofficially reserved for the Franklin High team. It was a patch of grass cluttered with our gym bags and warm-up clothes, where we did our stretches before the race and collapsed afterward. Tilting my head back and taking big gulps of air, I staggered toward my teammates.

Elena was already there, red-faced and panting, but still so untouchably beautiful. A gaggle of admiring Junior Varsity girls surrounded her as she opened her bag and pulled out a blue sweatshirt. And standing in front of the pile of bags was a heavy, balding man in a totally ridiculous tracksuit, colored a shade of orange so ugly it hurt my eyes to look at it.

That was Mr. Goldman, the coach of our boys and girls cross-country teams. He held a pencil in one hand and a clipboard in the other, but his eyes were fixed on me. He pressed his thick lips together as I came toward him. The skin under his left eye twitched.

"What happened, Joan?" His voice was extra annoying today. It was loaded with sarcasm, which Goldman seemed to think was a useful coaching tool. "Did you wake up on the wrong side of the bed this morning?"

I stood a couple of yards away from him. His breath smelled like Parmesan cheese, so I always tried to keep my distance. "It wasn't my fault." I was still breathing hard, still trembling. "I had to stop."

Goldman looked down at his clipboard. "Your time was twenty minutes, nineteen seconds. More than a minute slower than last week." He pointed at the digital clock next to the finish line. "In fact, it was your worst time of the whole season. By far."

He raised his voice a bit, obviously for the benefit of his audience. Because the Girls Varsity race was the last one of the day, all the other runners from our school—the JV girls and both of the boys teams—were hanging out by the pile of gym bags, shivering in their sweatshirts and waiting for the final results. They didn't stare at Goldman and me, but they were all listening in. They put their own conversations on hold and pretended to look at the finish line.

I had no interest in entertaining them. There were a lot of jerks at my school, stupid, jealous kids who spent all their time spreading rumors and idiotic videos on their phones. And even the decent, friendly kids had started acting weird around me after they heard what happened to my sister over the summer. So I turned away from Goldman. I went to the pile of bags and bent over, hunting for mine. It was purple and marked with a round sticker that said DISSENT IS PATRIOTIC.

The coach wasn't done, though. He stepped sideways, moving between me and the bags. "I don't get it. You were running so well the past few weeks. You could've easily won this thing." He shook his head. "I'm disappointed, Joan. This was the last race of the season, the last high-school race you'll ever run, and it wasn't your best effort."

I grimaced. Goldman was such a creep. In addition to being our cross-country coach, he taught calculus at Franklin High, and he was still mad at me for something that happened way back in my freshman year. Because I was so good at math, the school principal let me skip all the easy courses—algebra and geometry and trigonometry—and go straight to Goldman's calculus class. But even his honors course was too easy for me, so I asked the principal if I could take the advanced math courses at City College, which is right next to Franklin on 140th Street. The college has a special arrangement with our high school, allowing us to use their facilities. Taking classes there was the obvious solution for me, and the principal had no problem with it, but Goldman got insulted. He took his revenge by torturing me during cross-country practices, criticizing everything I did. It was so petty and pathetic.

But the season was over now, and I didn't have to take it anymore. I stood up straight and looked him in the eye. "I told you, I had to stop running. There was a boy lying on the ground, a kid from another school. I thought he was hurt, so I stopped to help him out, okay?"

Goldman raised one of his eyebrows. "Hurt? Did he break his leg or something?"

"Well, it turned out he wasn't hurt at all. But when I first saw him—"

"So instead of running the race, you were playing doctor with somebody?" He snorted, and I got a whiff of his Parmesan breath. "Sorry, I don't buy it. You're making excuses."

Seriously, I wanted to punch the guy. I stepped backward and glanced at my teammates, hoping someone would stick up for me or at least show some disgust at the coach's behavior. But all the other kids just stood there, stone-faced and silent. They were afraid of Goldman. He was a notoriously tough grader, and Franklin High was a school for grade grubbers and brown-nosers. Everyone except me had to take Goldman's calculus course in their senior year.

I focused on Elena. She'd put on her sweatshirt, which fit her perfectly of course. "Hey, Elle, you saw that kid, right? The boy I was trying to help?"

She looked over her shoulder at me, her black hair dangling to the side. Until a month ago, Elena and I had been best friends. We lived on the same block—78th Street and Broadway, on Manhattan's Upper West Side—and we'd gone to the same elementary and middle schools. We both joined Franklin's cross-country team in our freshman year and became co-captains two years later. But I did something dumb on the first day of school this year, and Elena started avoiding me. I hadn't had a real conversation with her in weeks.

Now she stared at me for a few seconds, frowning slightly. Then she shook her head. "Uh, no, actually. I didn't see anything."

My mouth fell open. I couldn't believe it. "But you ran right past us."

Elena shrugged. "Sorry. I didn't see any kid on the ground, or you either. You ran way ahead at the start of the race, and I lost sight of you. I didn't see you again until the end."

I closed my mouth and gritted my teeth. *Why is she lying? Does she really hate me that much?* It took all of my willpower to keep my voice steady. "How can you say you didn't see me? You *passed* me! You were just a couple of feet away!"

Goldman stepped forward and held out his hands like a traffic cop. "All right, cut it out. I don't care about your excuses, Joan. I'm just disappointed that you didn't give it your all today. The Varsity team had a real shot at the city championship, but now it's not gonna happen. You let everyone down."

I'd had enough. Looking past Goldman, I spotted my purple gym bag at the edge of the pile. I grabbed it and marched away from the team, glaring at everyone in sight. They were a bunch of freakin' cowards. I hated them all.

I headed back to the finish line. The last of the Girls Varsity runners were coming in now, the stragglers from Queens and Staten Island. A couple of race officials sat at a folding table and logged the results, calculating each high school's score from the finishing positions of its five fastest runners. An ancient gray-haired guy penciled numbers on a clipboard while a slightly younger guy tapped the keys on a calculator. The sky had turned cloudy, and a strong, cold wind ruffled the pages attached to the clipboard. The race officials seemed a little frazzled, but I approached them anyway.

"Excuse me? I'm trying to find one of the boys in the Junior Varsity race?"

The younger guy looked up from his calculator. "We're a little busy right now. Can you—"

"The thing is, I'm kind of worried about him. He fell down on the course and didn't finish with the other JV kids. I think he might still be out there."

The older guy looked up too and studied me for a moment. Then he started leafing through the pages on his clipboard. When he found the page he wanted, he squinted at the list of names there. "All the JV boys finished the race. The last one had a time of 31 minutes and 54 seconds."

I leaned over the table and craned my neck, trying to get a better look at the list. "The boy's race number is 137. When did he—"

"137? That can't be right."

"Believe me, I'm sure of it. I have a great memory when it comes to numbers."

The old man shook his head. "There's no number below 200 in the city championship race. The race numbers of the girls go from 200 to 550, and the boys' numbers go from 600 to 999."

The cold wind blew again, strong enough this time to rip one of the pages out of the old man's clipboard. The younger guy yelled, "Crap!" and ran after it.

I backed away from the table. The sky was getting dark, threatening rain. I looked up at the gray clouds, then closed my eyes and shivered. I remembered what the boy had said as he lay on the muddy ground.

The end of all things is at hand.

Chapter Three

After the race, I went to the Burger King across the street from Van Cortlandt Park. I didn't go straight home because I knew I'd run into the kids from my team if I went to the subway station at 242nd Street. But no one from Franklin High would ever step foot inside a Burger King, not in a million years. They were high-achieving students from high-achieving families who followed the number-one rule of Good Helicopter Parenting: Never let your kids eat fast-food. The rule had been carved so deeply into the kids' brains that they probably got nauseous if they even looked at a Whopper.

I bought a Whopper and a large order of fries and found a table at the back of the restaurant. I was usually mad hungry after a cross-country race, but this time I felt sick after a few bites of hamburger. I couldn't finish the rest, so I just sat there and looked out the window. It started raining outside, an epic downpour. Hundreds of kids hurried out of the park and dashed for the subway station.

I waited about half an hour, until I was certain that everyone on my team had gone home. Then I headed for the station myself and got on the downtown 1 train.

It was a 45-minute subway ride from Van Cortlandt Park to my neighborhood. The first part of the ride wasn't so bad; the train line ran aboveground in the Bronx, and I could stare at the apartment buildings on both sides of the elevated tracks and watch the rainstorm taper off. But then the train crossed the bridge into Manhattan and rumbled into the subway tunnel beneath Broadway, and the train car's windows went black. I had nothing to look at, nothing to distract myself from all the things I didn't want to think about.

So I reached for my gym bag and fumbled inside it until I found my math notebook. That fall I was taking a new math class at City College, something called a tutorial, and it was completely different from all the college courses I'd taken so far. The professor, Dr. Laura Taylor, was an expert on number theory; specifically, she focused on the rational numbers, the ones that can be written as fractions. What made her course so different was that Professor Taylor said I could choose whatever kind of math I wanted to study. All I had to do was find an interesting problem and try to solve it. So that's what I'd been doing for the past few weeks, scribbling equations and theorems in a spiral notebook with the word TUTORIAL scrawled on its blue cover.

I loved that class. It was the only thing that made school bearable.

I opened the notebook and looked over what I'd written the night before. I've been a math nut ever since I was five years old, when Dad taught me how to add and subtract and multiply. A year later he started teaching me algebra, and by third grade I was teaching myself geometry and trigonometry. I went to the library and checked out all their books of mathematical puzzles, and I always kept a math notebook close at hand so I could jot down my solutions to the problems I was working on. I did the same thing in high school too, but now my notebooks were full of more complicated scribbles: differential equations and symmetry groups and matrices and vector spaces.

I examined my latest work as the train sped downtown. I got so involved in the equations that I almost missed my stop at 79th Street. Seriously, I lost all sense of time and place. When I got off the train, I hardly noticed the musty smell of the station or the *clank-chunk* of the turnstiles or the damp steps of the stairway that led up to the street. I ignored the traffic on Broadway and the crowd of people at the bus stop and the parade of dog walkers and window shoppers on the sidewalk. I looked above the chaos of the street and pictured a number line in the sky, endlessly long and perfectly straight, stretching from negative infinity on one side to positive infinity on the other. It was like a rainbow, but more precise, more beautiful.

The good feeling didn't last, though. It faded as I got closer to my home, an old six-story apartment building on 78th Street. It was the black sheep of the Upper West Side, a shabby red-brick rental building surrounded by all the gleaming condominium palaces where the neighborhood's rich people lived. I walked into our lobby and got on the elevator, which made alarming noises as it crept upward.

But the worst part was opening the front door to our apartment, because I knew my parents would be home. Life with them had become a challenge since Samantha died.

Mom was on the living-room couch, talking on her cell phone. She looked up at me and waved hello as I stepped inside, but she didn't pause her conversation, not even for a second.

"Yes, definitely…the work we do at Harvest Team follows all the principles and values you're talking about…when you're in New York next week you should really take a look at…yes, yes, you can bring your stepdaughter too. I'd be happy to set it up."

Even though it was Saturday, Mom wore her work clothes: white blouse, red jacket, plaid skirt, nylons. Her laptop was on the coffee table, and she hunched over its keyboard as she talked on the phone, doing two things at once. I turned away from her and dropped my gym bag on the rug and tore off my sweatshirt. Then I headed for the kitchen. My stomach still felt queasy, but maybe I could eat a piece of toast.

I used to get mad at Mom when she did work on the weekends, but now I didn't care. It was a relief, actually. Work was therapy for her. When she wasn't working, she'd get nervous and depressed, basically torturing herself over what happened to Samantha. Even worse, she'd focus her attention on me and start asking a million questions, trying to probe my feelings—*How is school? Have you talked to Elena lately? Are you getting along better with the other kids?* She was so worried about my mental stability that I sometimes felt like I was living in an asylum and she was my nurse. I told her at least once a day that I was doing fine, that I didn't need a psychologist, and that I'd immediately let her know if my emotional state began to teeter. In response she just narrowed her eyes, looking for symptoms, signs of trouble. I was under observation 24/7, and it was exhausting for both of us.

But I'd gotten a bit of a break for the past few days, because Mom was in the middle of her "grant season," the busiest time of the year for her. She worked as a fundraiser for a nonprofit organization called Harvest Team, which donated food to dozens of soup kitchens and food pantries across New York City. She'd devoted her career to helping the poor and homeless, but the ironic thing was that she spent almost all her time with rich people, persuading them to make donations or provide grants. That's why she'd dressed for work on a Saturday—she'd probably gone to a fundraising luncheon earlier that afternoon and was headed for a cocktail party or charity auction later that evening.

I grabbed a loaf of bread from our freezer and put a couple of slices in the toaster. While I waited for the toast to pop up, Mom finished her phone call with a loud, "Okay, I'll get in touch with you on Monday!" Then she rushed into the kitchen, grinning and manic. "Oh my God, Joanie, you won't believe who I was just talking to!"

She leaned close to me and brushed her hair from her eyes. Mom was pretty. She had great skin, young-looking and unblemished, and her hair was stylishly colored with blond streaks. The excitement on her face made her look even younger, and I felt a twinge of jealousy. Frowning, I stared at the toaster. "Where's Dad? Is he home?"

"Yeah, yeah, he's at his desk. He's still writing that magazine article he should've handed in yesterday." She waved her hand in the direction of their bedroom, brushing away my question just like she'd brushed her multicolored hair. "But listen, I have some news. That was my cousin Teresa on the phone. You know, the hippie cousin? The one who ran away from home?"

Mom came from a big, crazy Catholic family, full of scolding aunts and embarrassing uncles, Italian and Irish, who'd mostly given up on their religion but still went to church once a year on Easter Sunday. Although I had a vague memory of Mom's stories about cousin Teresa, I wasn't interested in talking about her right now. Luckily, the toast popped up at that moment, giving me something to do instead of answering her question.

"Come on, Joanie, I told you all about her. She's the New Age girl, the one who ran off to New Mexico to make bead necklaces and study astrology." Mom leaned a little closer and lowered her voice, as if she were about to tell me a world-shattering secret. "Well, get this: after living in the desert for sixteen years, Teresa married an Internet millionaire, and he gave her enough money to start a charitable foundation. And now she's coming to New York to decide which nonprofits she'll give donations to, and one of the organizations she's considering is Harvest Team. Can you believe it?"

I went back to the fridge, pulled out a stick of butter, and started spreading it on my toast. I knew I was being kind of mean to Mom by not responding to her news. I should've at least pretended to show some interest, just to be polite. But I also knew my own tendencies, and when I got into a mood like the one I was in then—tired, sad, confused—the best strategy was to keep my mouth shut. If I opened it, I had no idea what would come out.

Mom stared at me. My silence alarmed her, so she gave me a once-over, studying me from head to foot to see what was wrong. Her eyes swept over my track shorts and sneakers, and an instant later she smacked her hand against her forehead.

"Oh God. You just came back from your race, right? And I didn't even ask you about it." She shook her head. "I'm sorry, babe. How was it? How did you do?"

I took a bite of my toast. Mom probably thought I was getting back at her by deliberately keeping her waiting, but that wasn't it. I needed a moment to organize my thoughts. I didn't know where to start.

I decided to stick with the facts. "I came in eighth. My time was twenty minutes, nineteen seconds."

Mom opened her mouth wide. "That's fantastic! Eighth place, out of all the runners in the city! You should be so proud of yourself."

She was faking it. The phoniness was so obvious, it wouldn't have fooled a three-year-old. I'd told Mom the night before that I was hoping to finish the race in the top five, maybe even come in first. So she knew I was disappointed, and now she was just trying to cheer me up.

"No, it wasn't fantastic. I should've done better. The coach said I let everyone down."

Mom furrowed her brow. A pair of vertical lines appeared above the bridge of her nose, and her eyebrows curved like scimitars. "Goldman said that? Seriously?"

"Well, if I'd run as fast today as I ran last week, I would've come in first. And our Varsity team would've been city champs. So Mr. Goldman was disappointed, and he let me know it."

"That…that's outrageous!" Mom clenched her right hand into a fist. She slammed it on the kitchen counter, and the toaster rattled. "It's unbelievable! Goldman knows your whole story, he knows what you're dealing with right now, and yet he goes ahead and says something so…so…"

Her voice trailed off and her face reddened. I'll say this for Mom, she had more than her fair share of the protective Mama Bear instincts. If Mr. Goldman had stepped into our kitchen at that moment, she would've bashed his head in with a skillet.

She raised her fist, ready to strike again. "You know what I'm gonna do? I'm gonna go to your school Monday morning and go straight to his classroom.

No, wait, I have a better idea. I'll go to the principal's office and tell her exactly what Goldman said."

I dropped my toast on my plate. I was too nauseous to eat it anyway. "No, you're not gonna do that."

"He should have some sensitivity, for God's sake! He can't—"

"Look, you're missing the point, okay?" I raised my voice, almost yelling at her. I had to do that with Mom sometimes. It was the only way to get her attention. "The problem isn't Goldman. He's just doing what stupid coaches do. That's not the thing that's bothering me."

Mom stepped backward. She unclenched her hand and took a deep breath, then another. After a couple of seconds her face relaxed, smoothing her forehead. Then she looked me in the eye. "Okay. I'm sorry, Joanie. Tell me what's bothering you."

"No, forget it. It's just a thing that happened during the race. It doesn't matter."

"What happened? Come on, you can tell me."

I wanted to tell her. And if the situation between Mom and me had been a little more normal, maybe I would've. I would've explained why I'd stopped in the middle of the race. I would've told her about the boy I saw in the woods, the kid who looked dead but wasn't. I would've described how he wore a magic number on his chest, and how he seemed to be invisible to everyone else. And I would've repeated the words he'd spoken in that deep, gigantic voice.

But I couldn't. I'd experienced something strange, something I couldn't explain. I knew it would frighten my parents just as much as it had frightened me.

We weren't a religious family. We didn't believe in ghosts or supernatural visions. We were modern, liberal, forward-thinking New Yorkers. If I'd told Mom everything I'd seen and heard, she would've smiled and nodded, and then she would've taken her cell phone into the bathroom and called the psychiatrists. She would've pulled me out of school and put me in a treatment center for at-risk adolescents, probably a fancy clinic somewhere in New England. I'd have to take a million psychiatric tests and attend group-therapy sessions and go on long therapeutic hikes with the other screwed-up teens. Worst of all, I'd have to meet every morning with the clinic's oh-so-sympathetic grief counselor, who would force me to talk about Samantha.

Which I just couldn't do. I'd loved Sammy too much. And I hated the world for killing her. I had so much anger, but there was no one to blame, nothing to punch or scream at. All I could do was cry, and crying didn't help, not one bit.

So I shook my head and stepped around Mom. I stormed out of the kitchen and stomped into my bedroom. And to make my feelings clear to everyone in the apartment building, I slammed the door behind me.

Chapter Four

The apartment got very quiet after I threw myself on my bed. A few seconds later, though, I heard furious whispers coming from the hallway outside my room. Mom and Dad were arguing in their bedroom, which was across the hall from mine, but they kept their voices low, most likely because they were arguing about *me*.

Their marriage was in trouble. I could tell from the long, tense silences between them, and all the weekends and evenings they spent in separate rooms. Mom usually hung out in the living room, either watching TV or chatting on the phone with her friends, while Dad holed up in their bedroom, playing dopey games on his laptop instead of working on the magazine articles he was supposed to write. They'd gotten into this habit years ago, but after Samantha died it got worse. On the rare occasions when Mom and Dad crossed paths—like when we all agreed to have a family dinner in the dining room—they couldn't even engage in polite conversation anymore. They made snarky comments. They interrupted each other. They ate their food as fast as they could, then retreated to their rooms.

It was kind of like the Cold War between America and Russia, which I'd studied in U.S. History during junior year. My parents had plenty of arguments but no screaming matches. They held themselves back from all-out war, because they knew it would blow up their whole world.

After five minutes, the whispering stopped and I heard the rapid steps of Mom's high heels across the living room, followed by the heavy slam of our front door. She'd obviously left the apartment to go to the charity auction or cocktail party she'd dressed up for. Then, after another five minutes, I heard a soft knock on my bedroom door.

"Joanie? Can I come in and ask you a question?"

Dad's voice was tentative. Mom had probably ordered him to get off his computer and find out what was wrong with me, but he wasn't suited for this kind of assignment, so in all likelihood he was hoping I'd tell him to go away. Instead, I asked, "A question about what?"

"Uh, complex manifolds?"

I sat up in bed and stared at the closed door. "Is this a trick to get me talking? Because Mom's worried about me?"

"Yeah, it is. But it's also true that I just read something on Wikipedia about complex manifolds, and it totally confused me. I think they're geometric structures or something?"

I sighed. Dad was a freelance journalist who liked to write articles about science, and lately he'd taken an interest in physics and astronomy. But understanding those subjects requires some math skills, and Dad's education in math had never progressed beyond high-school trigonometry. So he'd started coming to me for help whenever something stumped him.

And here's the funny thing: I didn't mind helping him. The way I felt about math was like how baseball fans feel about the World Series, or how priests feel about Jesus. I loved thinking about it, talking about it, making jokes about it. It was so beautiful and true, I wanted to share it with everyone.

"Okay. Come in."

Dad opened the door and stepped inside. He was short and gray-haired, not nearly as good-looking as Mom, but he had nice brown eyes and a good sense of humor. He wasn't a normal dad—he was a misfit, half-Jewish and half-Unitarian, an aging nerd who still liked punk rock and comic books. He didn't have an office job, so he hardly ever shaved or wore regular adult clothes. Right now, for instance, he wore khaki shorts and a New York Mets T-shirt and a pair of ragged flip-flops. He carried his laptop into my room and took a seat on the beanbag chair next to my bed.

"Take a look at this." He opened his laptop and showed me the Wikipedia page that had confused him. "It makes no sense at all. Just look at this jargon, all these bizarre terms—K3 surfaces, Calabi-Yau manifolds. I mean, *Calabi-Yau*? It sounds like something out of a Dr. Seuss book, right?"

I sat on the edge of my bed and looked at the computer screen. Then I reached over Dad's shoulder and closed his laptop. "Start at the beginning. First, tell me what you're writing."

"It's an article for *Popular Science* about physics. Specifically, string theory. You've heard of it, right? Did they mention string theory in that physics course you took last year?"

I nodded. I'd taken Advanced Placement physics in my junior year, but I hadn't liked the class as much as I'd thought I would. The teacher was awful, and he'd forced us to do a ton of really lame experiments, like rolling a tennis ball down a ramp and watching a Slinky jiggle up and down. It was so boring and pointless, redoing all that kindergarten-level research. That's why I liked math so much more than science—mathematicians never had to do experiments. They never had to deal with the real world at all. Their work was purer, cleaner.

It was in that physics class, though, that I learned about the magic number. I don't know why, but it sank into my brain: the fine-structure constant, also known as alpha, which quantified the strength of the electromagnetic force between two charged particles. What made the number magical was that it seemed to be perfectly calibrated to allow the creation of stars, planets and life. If alpha were much higher or lower, atoms and molecules couldn't form and stars couldn't shine. In other words, the whole universe would be barren and lifeless if the value of alpha were much different from 0.00729735, which is weirdly close to the reciprocal of 137.

I turned away from Dad and saw that number again in my mind's eye. I pictured it on the track shirt of the boy I'd seen in Van Cortlandt Park. The number had followed me home.

I blinked a few times to get the image out of my head. Then I turned back to Dad. "Yeah, the physics teacher mentioned string theory, but he didn't say much about it. It's a new kind of physics, right? They're trying to explain all the laws of nature, why the universe is expanding, that kind of thing?"

Dad smiled, excited. "Right, exactly. These crazy physicists have been wrestling with the problem for forty years, and they still haven't come close to solving it. *Popular Science* wants me to write an update about the latest progress the scientists have made. But I won't feel comfortable writing about the subject until I understand the basics." He reopened his laptop and typed in a different Web page. "So I found a research paper by Edward Witten. He's a Princeton physicist, one of the top thinkers in string theory. But the math in his paper is ridiculously hard. I mean, can you make any sense of this?"

I looked over his shoulder. His computer screen was packed with equations, from the top of the page to the very bottom. The font was so small, I had to lean closer and squint.

Dad shook his head. "And this is just one page. There are fifty-three more just like it."

It was a real mish-mash of mathematics: differential equations and matrices and integrals and tensors. I'd taken half a dozen math courses at City College, so I recognized most of the terms and symbols, but this stuff was a lot more complicated than anything in my textbooks. I grabbed the laptop out of Dad's hands so I could get a better look at the screen.

He waited half a minute, and then his impatience got the better of him. "So? What do you see? Any insights from the teenage math genius?"

I pointed at the screen. "I'm not sure. This is difficult stuff."

"Oh. That's bad news." Dad raised his hand to his forehead. "If it's difficult for you, then it's going to be *impossible* for me."

"The math is interesting, though. It seems to be a description of a topological space." I tapped on the equations at the bottom of the screen. "The space has a vanishing first Chern class and a Ricci-flat metric."

"Okay, now you lost me. I have no idea what you're talking about."

"Hey, I don't completely get it either." I touched the laptop's trackpad and scrolled down, trying to follow the equations. "But if you want, I can read the whole paper. Then maybe I could give you a better explanation."

Dad started massaging his forehead. At the same time, he let out a tired sigh. "No, it's not worth it. The article was due yesterday. I should just go ahead and start writing it."

"Are you sure?"

"I appreciate the offer, Joanie, but I don't want to make you miserable too." He reached for his laptop and took it back. "I'm just not smart enough for this kind of assignment. I should've never agreed to do it."

He ran his hand through his gray hair, pulling it back and showing how much it had receded. Dad looked old today, even older than usual. In his younger days he was an editor for *Newsweek*, but people stopped buying that magazine, so he got laid off. He became a freelancer, writing articles for several magazines and websites, and he was happy for a while, even though he wasn't making as much money as he used to. But he seemed to lose faith in himself after Samantha died. In the past few months he hadn't gotten many writing assignments, and he'd started handing them in late. And that became another source of tension between him and Mom. She kept telling him to snap out of it and get back on his feet. But he couldn't.

Dad closed his laptop, but he didn't get up from the beanbag chair. He just sat there and stared straight ahead at my closet door, which was covered with a poster that said I AM WOMAN, HEAR ME ROAR. He had a vacant look on his face—mouth half-open, eyelids drooping. I wanted to ask him if he was okay, but I was afraid that would just make him get up and leave the room. So I tried to think of a better question.

"I bet you can write a good article even if you don't understand the math. You interviewed some of the physicists who work on string theory, right?"

He nodded. "Yeah. Eight of them, including Witten. Mostly from Princeton and Harvard."

"Well, what did they say? Did they explain what they're trying to do?"

He turned away from the closet, but he still had that vacant look. "You know what? I think the physicists are depressed. When they started working on this string-theory problem, way back in the Seventies, they had a lot of hope. They thought they could come up with a Theory of Everything that would explain all the forces and particles in the universe—gravity and atoms and electrons and galaxies, the whole shebang." Dad waved his hand in a big circle. "The scientists expected too much, see? And now they're disappointed. Their big idea didn't work, and they don't know what to do next."

"So they've made no progress? The whole thing is a bust?"

"No, not exactly. The physicists still love their theory. They say it's a thing of beauty, although you can't really appreciate it unless you understand the math. But it didn't live up to their hopes. The theory turned out to be this wild, monstrous thing that didn't explain our universe at all. Instead, it split up into a gazillion different theories, and each one predicted different things, a different universe. It's a mess, Joanie, a screwed-up cosmic mess." He frowned. "And you know what's the worst part? You know why the scientists think they failed? Because they now believe it's impossible to explain the universe. Their theory predicts that a gazillion possible realities can exist. And that means the theory is useless, because it explains nothing."

Dad was breathing hard by the time he finished, the air hissing between his teeth. He was upset about this, *really* upset. It surprised me to see him that way, and it scared me a little too. I needed to calm him down, get him back to normal.

"Okay, listen. I think you can write this article pretty easily. All you have to do is write down the things you just told me. And maybe add a few introductory paragraphs. And some quotes from the scientists."

"No." He shook his head so vigorously, his thin hair flapped above his scalp. "I'd rather just forget it."

"I mean, it's not a success story, but that doesn't mean you can't write about it. I think it could still be an interesting article if you—"

"No!" Dad shot to his feet, leaving the laptop on the carpet. "I can't do it!"

I stood up too, next to my bed, staring at Dad in disbelief. His face had turned red, especially his cheeks, and his jaw muscles quivered. He scowled and pointed at me, his finger trembling.

"I don't want to write the story! I'm not interested in it anymore!"

I held my hands up, surrendering. I didn't know what else to do. "Okay, okay! Just—"

"What's the point of life if there's no explanation? If the reason for the universe is hidden *forever*?"

His voice cracked on the last word. He took a step backward, then another, backing away from me until his shoulders pressed against the wall. Then he slid down to a crouch, sinking to the floor.

I just stood there, swaying in the middle of my bedroom. I couldn't think, couldn't move. My chest tightened as I stared at Dad, because I'd seen him collapse like this once before, only four months ago. It had happened right here in my bedroom, while Dad was talking to me about another article he wanted to write. Mom had burst into the room, crying, and handed her cell phone to Dad. As he held it to his ear, he backed up against the wall and slid down to the floor.

That was June 28. The phone call had come from Nauta, a small, poor village in northeastern Peru. The caller was the principal of the village's school, where Samantha had gone to teach English for the summer.

Now it felt like the catastrophe was happening all over again, but without Mom and the cell phone this time. Dad had the same expression on his face: stunned, ashen. And just like before, I couldn't think, couldn't move.

After a while, Dad struggled to his feet. He looked at me for a few seconds and bit his lower lip, but he was too embarrassed to say anything. So he turned away. He left my room without a word.

He closed the door behind him, and the apartment got quiet again. All I could hear was the distant siren of a fire truck, screaming down Broadway.

I understood what had just happened. Dad was still trying to make sense of Samantha's death, still searching for some meaning in it. That's why he'd wanted to write his article about the mathematical design of the universe. But instead of getting some solace from the physicists' equations, he'd glimpsed a horrible possibility: The universe might be incomprehensible, a cosmos without any plan or design. And if the laws of nature were totally random, there were no explanations for life or death, and it didn't even matter that Samantha was gone, because the world was racing toward extinction anyway. *Everything* would be lost.

The fire truck's siren grew louder. I clenched my hands as I listened, angry and upset, feeling the same frustration that Dad must've felt. I was a mathematician, so I hated disorder, and I couldn't stand the idea of an arbitrary universe. I wanted the world to have a logical set of rules. I thought of all the beautiful theorems I'd learned, the whole intricate web of algebra and geometry, and it seemed impossible that so much theoretical perfection could've arisen from a meaningless reality. I refused to believe it.

No. There's a reason for all this. And for what happened to Samantha. I just have to search for it. I owe her that much.

After a few more seconds, the siren faded. Then I knelt on the carpet and opened the laptop that Dad had left there. And I stared at the equations on the screen.

I made a promise to myself.

I was going to find an explanation.

gossip in the school—so I glowered at her, scrunching my face and baring my teeth. But by the time I made the face, she'd already turned away from me, so she never saw it.

"You should know a few things about Catharism. It was a dualist religion, which means it had two Gods instead of one: the good God who rules over the invisible spiritual world and the bad God who created the visible world we live in. The Cathars believed that all physical objects were tainted with sin, so it was holier to avoid material things. They lived in poverty and abstained from eating meat. And they rejected the authority of the Catholic Church and its priests. So you can see why the pope was so anxious to get rid of them."

I started taking notes. This stuff about Catharism was actually kind of interesting. I liked the idea that Christianity used to be more complicated. Medieval Christians couldn't even agree on the number of Gods. It took them hundreds of years to get their story straight.

But the subject made me a little uncomfortable too. It reminded me of the boy I saw in the woods, on the muddy trail in Van Cortlandt Park. And I didn't want to think about him again. I wanted to put that incident behind me and let it fade with time. In the two days since it happened, the memory of it had already grown a little fuzzy and unreal. Very soon, I hoped, it would become like a dream, an uncertain, unreliable, fantastical vision that I wouldn't have to believe anymore.

Then I felt a tug on the back of my T-shirt. Someone had just slipped a finger inside the shirt's neck hole, right below the split ends of my short ratty haircut. The finger pulled the neckline back for a second, very gently, then let go. One of my classmates was teasing me, trying to get my attention, trying to goad me into turning around.

I clenched my hands, enraged. Blood rushed to my head, and everything looked pinkish for a second. In an instant, I spun around in my chair, ready to deliver a medieval punishment to the sorry loser who'd just provoked me.

But in the seat behind me was Cynthia Chen, the smallest girl in the senior class. She was barely four-foot-eleven and so quiet and shy that I hadn't heard her speak more than five times since freshman year. She said something now, though, as I stared at her. She pointed at my shirt and whispered, "Inside-out."

I looked down at the shirt. There was a tag with laundry instructions a few inches below my armpit. Fabric seams looped over my shoulders and ran down both sides of my chest. It was a white shirt, and on the front of it was a feminist slogan in big red letters. I could see the words through the fabric, but backwards, as if I were looking at them in a mirror:

ИAMOW MA I

All the rage seeped out of me and curdled into embarrassment. Now I knew what Nate had been trying to tell me and what Kathy had laughed at. I'd overslept that morning, and in my rush to get dressed I'd put on my T-shirt inside-out. Now everyone at Franklin High would have yet another story to tell about Joan Cooper, the sad girl whose sister died, the lonely geek who couldn't even dress herself.

"So this was a religious war between two kinds of Christians, and it was very brutal, even compared with the other crusades. In July 1209 an army of crusaders attacked the city of Béziers, which was packed with twenty thousand people, about half of them Cathars and half Catholics. But when the crusaders rushed into the city, they killed everyone in sight. And when someone told the leader of the army that his men were slaughtering good Catholics along with the heretics, he replied, '*Caedite eos. Novit enim Dominus qui sunt eius.*' Which is Latin for 'Kill them all and let God sort them out.'"

I continued listening to Mr. Owens, but now I was too depressed to take any more notes. I took my eyes off the man and let my gaze wander off to the side, to the seat at the far right of the classroom's front row. That's where Elena Gutierrez sat, hunched forward in her chair, her face just inches above the spiral notebook that she was filling with her perfect handwriting.

Elena took school very seriously. Her father was the super of an apartment building down the street from mine, and her family lived in the super's apartment on the building's ground floor. She planned to be the first woman in her family to go to college and get a Ph.D. Her favorite subjects were poetry and biology, and she'd told me many times that her dream was to become a cancer researcher. Two of her grandparents had died of cancer.

Even though she sat on the other side of the room, at least twenty feet away, I could see her so clearly. She wore a white cotton blouse with a V-neck collar, which was just right for school wear, not too plain or too fancy. Her jeans were a lot newer and bluer than mine, and her sneakers were cute Vans slip-ons. She had long slender arms and delicate hands, with no rings or fingernail polish, and her skin was smooth and tawny. Her long black hair was in a ponytail that morning, pulled away from her face, allowing me to stare at her brown eyes and furrowed brow. She pursed her lips in concentration as she listened to Owens.

"By 1229 the Cathars were defeated. The French king took firm control of the southern part of the country and ordered his vassals to hunt down any remaining heretics. The Cathars were forced to wear yellow crosses on their clothes as a sign of their repentance. And those who refused to repent were burned at the stake. That was the beginning of the Inquisition, which we'll talk about much more over the next few days."

I kept staring at Elena. I should've never kissed her.

Chapter Six

After History class, I ducked into the girls' bathroom, took off my T-shirt, and put it back on the right way. Then I sat through my English, Spanish, and Art classes, occasionally taking a few notes but mostly watching the clock. I was anxiously awaiting the only interesting segment of my schedule, which was the hour I spent every Monday afternoon with Professor Laura Taylor of City College's math department.

Toward the end of Art class, while I was supposed to be finishing a pencil drawing, I reached for my backpack instead and pulled out my math notebook. I held it open on my lap so that the Art teacher wouldn't see it. Then I leafed through the pages and looked over the equations I'd written the night before.

After my conversation with Dad on Saturday, I gave myself a crash course on string theory. First I read everything I could find on the Internet, then I went to the Columbia University bookstore and bought a couple of advanced physics textbooks. I got so excited about the mathematics that I hardly left my bedroom for the rest of the weekend. By Sunday night I'd thought of some new approaches to the theory, and I stayed up till 4 a.m. writing them down.

That's why I'd overslept on Monday morning. And put on my shirt inside-out. But it was all worth it, because now I was ready to show my ideas to Professor Taylor. She'd said I could work on any math problem that interested me, and now I knew what I wanted to do. The spirit of my sister—whatever that meant, wherever she'd gone—had inspired me.

At 12:20 p.m. the fifth-period buzzer finally rang. On Mondays I was allowed to leave the school building and go to City College's cafeteria for lunch; my routine was to buy a sandwich there, wolf it down, and then go straight to Professor Taylor's office, which was in a building across the street from Franklin High. But sometimes there was a long line at the cafeteria, and I really hated any delays that cut into my time with Laura. So as soon as the buzzer rang, I dashed out of the Art room and down the corridor, sprinting toward the school's front doors.

I was just fifteen feet from freedom when Principal Barnes came out of her office and stepped into my path. I broke stride, my sneakers squeaking against the linoleum, and lurched to a halt right in front of her. She gripped my shoulders to keep me from stumbling.

"Whoa! Slow down, young lady!"

A typical high-school principal would've chewed me out for running in the halls, but Crystal Barnes was far from typical. She was a lanky African-American in a snazzy black pantsuit, the kind of woman who was so consistently cheerful that it made you wonder if something was wrong with her. Smiling, she let go of my shoulders and looked me over.

"Hello, Joan. I knew you were in the City Championship race over the weekend, but I didn't realize you were still running it."

"Sorry, Ms. Barnes." Nervous, I shifted my weight from foot to foot. "I'm going to my tutorial with Professor Taylor, and I didn't want to be late."

"Well, I'm glad I caught you before you scooted off." She placed a hand on the small of my back and nudged me toward her office. "Let's make a deal, all right? If you'll do a little favor for me, I'll forget that you almost knocked me on my butt just now."

She led me into a small room crowded with stacks of textbooks. The principal's desk had a few knickknacks on it—a Rubik's cube, a model of an oxygen atom—and the wall behind it was plastered with posters bearing inspirational messages: BE LIKE A PROTON AND STAY POSITIVE and YOU HAVE TO BE ODD TO BE NUMBER ONE. There was a chair behind the desk and two more in front of it. Sitting in one of those chairs was a tall, pale boy with messy blond hair and bright blue eyes.

He stood up as Ms. Barnes and I stepped into the office. He wasn't a Franklin High kid. He was the right age to be in high school, probably sixteen or seventeen, but he was completely unlike *any* high-school student I'd ever seen. He was dressed way too formally, in a white shirt and a red tie and a dark blue suit that was several sizes too small for him. The sleeves of his jacket went only halfway down his forearms, and the hems of his pants dangled six inches above his shoes. He looked like he'd just gone through a growth spurt and hadn't had time to buy new clothes yet.

But the strangest thing about him was that he didn't seem embarrassed by his ridiculous outfit. He looked straight at me, dropped his hands to his sides, and bent forward at the waist. He actually *bowed* to me, as if we were dancers in a ballroom and a waltz was about to start.

The principal pointed at him. "This is Andrei Mishkin. From St. Petersburg, Russia. He and his family immigrated to America just two weeks ago." Barnes smiled at Andrei, then pointed at me. "And this is Joan Cooper, the pride of Franklin High. She's the math genius I was telling you about. She's already taken a bunch of math courses at City College."

Andrei nodded, his lips pressed together, his face very serious. Then he stepped forward and shook my hand. "Good afternoon, Miss Cooper. I am man."

His Russian accent was thick enough to confuse me. "Uh, I'm sorry, I don't—"

"It's a joke. *I am man.*" Grinning, he raised a hand to his chest and patted his shirt. Then he pointed at *my* shirt, specifically the feminist slogan on the front. "And you are woman. But I would've known that even if you didn't have 'I am woman' written on your clothes."

I frowned. I wasn't crazy about this guy's sense of humor. Maybe his joke would've been funnier in Russian, but I kind of doubted it. Before I could express this opinion, though, Ms. Barnes stepped between us.

"Andrei is a math genius too. He went to a special school in St. Petersburg and won all kinds of awards." The principal went to the chair behind her desk and sat down. "When his family came to New York and the school-district placement staff found out how smart he was, they decided the best fit for him would be here. Andrei's going to take most of his classes at Franklin, but our math courses would be too easy for him. So I'm trying to arrange some college courses for him, just like I did for you, Joan."

I looked closer at Andrei Mishkin, sizing him up. I knew, of course, that Russia was famous for its mathematicians. One of my heroes was Grigori Perelman, the Russian who'd proved the Poincaré Conjecture, a problem that had baffled mathematicians for almost a century until Perelman solved it in 2003. But he was a one-in-a-billion genius, and I seriously doubted that this gawky blond kid was on the same level. He had a wide, dopey grin on his face, like the painted smile of a clown. Three pink pimples made an isosceles triangle on his chin, and his hair coiled and twisted in every direction. He smelled funny too, like a mix of cabbage and mothballs.

Andrei turned to Ms. Barnes and bowed to her as well. "I appreciate your help, Madam Principal. My former school, St. Petersburg Lyceum 239, offered courses in differential equations, topology, and number theory, and I took advantage of them all. And now I look forward to attending the classes at City College."

I stared even harder at the guy. This whole European gentleman thing was probably an act. I suspected he was joking again, now at the expense of Ms. Barnes, but she didn't seem to care. She reached across her desk and picked up a handwritten note that had the Franklin High letterhead on it. "Well, I'm no math expert, so I don't know what would be the best courses for you. And we're seven weeks into the fall semester, so that might complicate things." She folded her note, slipped it into an envelope, and offered it to me. "Joan, could you give this to Professor Taylor? And please introduce Andrei to her? She can quiz him on his math skills and figure out what courses he should sign up for."

I didn't take the envelope. I was too surprised. "Wait, you want me to bring him to her office? Right now?"

"You're heading over there, right? Might as well kill two birds with one stone. You don't mind, do you?"

I *did* mind. I was dying to talk with Laura about string theory, and the subject was so complicated that I would need at least an hour to explain my ideas. The last thing I wanted was a weird Russian prodigy distracting us.

But I couldn't say no to Ms. Barnes. She was the nicest person at school, by far. She'd proved her kindness beyond a doubt on the first day of my senior year; while everyone else kept their distance and whispered behind their hands, she came right up and hugged me. She was the only one who said the right things and didn't make me feel like crap. I owed her.

I took the envelope. "No problem. I'll take care of it."

Andrei bowed again to the principal, and then we left her office. As soon as we reached the hallway, I gripped the too-short sleeve of his jacket and pulled him along.

"Okay, let's hurry. We're gonna get some lunch first."

* * *

Three minutes later, we stood in line at the sandwich station in City College's cafeteria. There were seven people ahead of us, which really sucked. I wondered for a second if we should get salads or burgers instead, but the line at the salad station was even longer, and the cafeteria's hamburgers were disgusting. So I fumed and fidgeted and stared at the clock on the wall, its minute hand slicing into my hour with Professor Taylor.

Andrei Mishkin didn't seem upset, though, not in the least. He was all smiles, turning his head left and right so he could observe the whole room, the college students cradling their lunch trays, the scowling, white-haired counterman making the sandwiches, the janitor mopping up a puddle of spilled soup. But what really got Andrei's attention was the cafeteria's carving station, where a blank-faced man in a chef's hat was cutting slabs of white meat from a large roasted turkey. Andrei let out a squeaky laugh, which was totally embarrassing.

"Look at that!" He pointed at the carver. "Look at the size of that bird!"

Everyone in the line stared at him. The City College students didn't like it when the Franklin High kids used their cafeteria; they thought we were too loud and childish and annoying, and right now Andrei wasn't helping our case. I spoke in a low voice, hoping he would get the hint and do the same. "It's a turkey. You don't have turkeys in Russia?"

"Yes, we do, but nothing like that one. It's as big as a beach ball!" He nodded vigorously and didn't lower his voice at all. "You have so much food here, so many different kinds. At Lyceum 239, we had only two choices for lunch, either sausage or porridge. And the teachers gave us only fifteen minutes to eat."

"Uh, Andrei? You don't have to shout. I can hear you just fine."

He raised his hand to his mouth, and a sheepish look spread across his face. "Oh! My apologies!" His voice was softer now, but still not quite discreet. "Please excuse me. Ever since I arrived in this country, I've been in a state of bewilderment. And the result is that I behave like a fool sometimes."

Andrei gave me a wink, slow and exaggerated, as if to demonstrate how foolish he was. That's when I saw him clearly for the first time. Until that moment, I'd assumed he was a jerk, an annoying know-it-all who enjoyed embarrassing me, but now I realized there was nothing mean or malicious about him. Andrei wasn't playacting. He really was as naïve as he seemed.

So I stopped frowning at him. I resolved to be nicer, more patient. "Why did your school give you only fifteen minutes for lunch? That's insane."

"Yes, we had to chew very quickly. But at my old school the rules were strict. The teachers wanted us to spend as much time as possible in class, because there was so much to teach us. Did you know that Lyceum 239 was named the best school in Russia for three years in a row?"

"No, I didn't know that."

"And our alumni include many famous mathematicians. Have you heard of Yuri Matiyasevich? Or Grigori Perelman?"

I jumped when I heard that second name. "Perelman? Of course I've heard of him!"

"He attended our school from 1981 to 1983. Are you familiar with Perelman's work in topology? How he used the Ricci-flow deformation technique on three-dimensional spaces to prove the Poincaré Conjecture?"

Andrei's voice was rising again, but now I didn't care, because we were talking about my hero, the most brilliant mathematician in the world. "Oh my God, yes! I read his whole proof, it's a freakin' work of art!"

"I agree with you, Joan, one hundred percent. The proof was very beautiful. The Clay Mathematics Institute awarded Perelman a one-million-dollar Millennium Prize for that work, you know. But he rejected the award."

"Yeah, I heard that too. It's weird, right? Why did he turn down the money? He totally deserved it."

Andrei shrugged. “He’s a strange man, even by Russian standards. He stopped talking with other mathematicians ten years ago, and now nobody knows what he’s working on. He’s unemployed and lives with his mother in a small apartment in St. Petersburg.”

“Seriously? He still lives with his mother? That’s—”

“Hey, you two! You’re next!” The interruption came from the scowling counterman. Andrei and I had moved to the head of the sandwich line. “Stop jabbering and tell me what you want!”

I ordered my usual Monday lunch—pastrami on rye, with extra mustard—and Andrei enthusiastically ordered the same thing. The counterman wrapped up the sandwiches for us, and we rushed over to the cash register. Then we left the cafeteria and speed-walked across campus, scarfing down our sandwiches as we headed for the Math Department building.

But on the sidewalk in front of the building, Andrei halted in mid-stride. I looked over my shoulder at him. “Come on. We’re almost there.”

He’d stopped smiling. He turned his head to the left and stood there wide-eyed and open-mouthed, like someone who’d just seen a car crash. In alarm, I turned in the same direction but saw no cars on Convent Avenue. There were no people either, except for a tall, broad-shouldered man in a long gray coat who was striding down the sidewalk on the other side of the street. I couldn’t see the man’s face—he was walking away from us—and in an instant he rounded the street corner and marched out of sight.

I turned back to Andrei, who was still staring at the corner where the tall man had disappeared. Then Andrei noticed I was looking at him. He tried to smile, but it was a feeble attempt. “I’m so sorry. For a moment I thought I left my wallet at the cafeteria. But it’s right here!” He slapped the pocket of his suit jacket.

I didn’t believe him. Not for a second. And I wasn’t polite enough to pretend I did. “What’s wrong? Why were you staring at that man?”

“Oh, no, I wasn’t staring.” Andrei shook his head. He did it too strenuously and too many times. “I’ve never seen him before.”

I stepped closer. “Listen, Andrei, you seem like a decent guy. But if we’re gonna be friends, we should be honest with each other.”

Wincing, he brushed the tangled hair from his eyes. He looked so nervous, I thought he’d bolt any second. “I *am* being honest with you, Joan. Really.” His voice dropped low, almost to a whisper. “And I want to be your friend, very much so. I had to leave all my old friends behind in Russia, so now I’d like to make new ones.”

He gave me a pleading look, tilting his head and raising his eyebrows. He was begging me to stop asking questions. And though I didn’t want to, I nodded. Andrei had recognized the man in the gray coat and was clearly afraid of him. But he wasn't going to talk about it, at least not now. I would have to wait until he was ready.

I reached for his shoulder and squeezed it. "Okay, we're friends. It's official."

He looked up, surprised. "Yes? Really?"

"Why not? But we need to talk about your clothes." I nudged him toward the entrance to the Math building. "You need to dress more casual. Get rid of the suit and tie. I'm speaking as your friend now."

Then we stepped into the building and climbed the stairs to Professor Taylor's office.

Chapter Seven

Like most of the City College campus, the Math building was more than a hundred years old and made of gray stone. It looked like a castle owned by a medieval English lord who enjoyed torturing his peasants. But the inside of the building was renovated and modern, and Laura Taylor's office was truly awesome. It was exactly the kind of office I wanted to have someday.

The room was elegant, uncluttered. The walls were bare. Although Laura had hundreds of books on every kind of math you can think of, they were all neatly shelved in a blond-wood bookcase. But the best thing was her desk, which was custom-made and shaped like an obtuse triangle. It had a vertex angle of 120 degrees, which meant that the length of the desk's longest side was equal to the length of one of the short sides multiplied by the square root of three.

Cool, right?

But the desk's shape was also practical. Laura had positioned an office chair behind each of the triangle's short sides and a whacking-huge computer screen on the longest side. It was the perfect arrangement for our weekly tutorials—when Laura and I sat in those chairs, we could look at each other and the computer at the same time, pointing at the equations and graphs on the screen as we discussed the math problems I was working on.

There was no chair for Andrei Mishkin, though. When I brought him into Professor Taylor's office, she turned away from her screen and gave us a puzzled look behind her glasses. She was petite and graceful, in her late forties or early fifties, with salt-and-pepper hair in a chin-length bob cut. She liked to wear eccentric clothing, and that afternoon she was in a flowered dress that looked like something Eleanor Roosevelt might've worn during the Great

Depression. After a moment she took off her glasses and wiped the lenses with a handkerchief.

"You've brought a friend with you today, Joan?"

Laura was formal and didn't have much of a sense of humor. I admired her as a mathematician—she was one of the best at City College—but I never felt relaxed around her. After seven weeks of tutorials, I still felt like I was trying to break the ice. I approached her desk and opened my backpack and pulled out the envelope that Principal Barnes had given me. At the same time, I pointed at Andrei.

"He's a new student. From Russia, believe it or not. He went to the same school as Grigori Perelman." I handed her the envelope. "Ms. Barnes explains it all in this letter."

Andrei stepped forward and bowed to Professor Taylor, but he didn't say anything. Laura had that effect on people, making them feel uncertain, cowed. She opened the envelope and quickly read the principal's letter. Then she shook her head.

"Unfortunately, I'm pressed for time today. I won't be able to do the tutorial *and* test Mr. Mishkin's math skills." She turned to Andrei. "Can you come back tomorrow at lunchtime? Then I can evaluate you and determine which of the math department's courses you should take."

Andrei nodded. "Yes, professor." He bowed again for good measure. "I'd be happy to."

"In the meantime, you're welcome to observe the tutorial. It'll give you a sense of how we do things here. But I don't have an extra chair, so I'm afraid you'll have to stand."

"Oh, I don't mind standing." He grinned and slapped his thighs. "I have strong Russian legs!"

Laura squinted at him, probably wondering the same thing I'd wondered, whether Andrei was making fun of her. Then she put her glasses back on and turned to me. "All right, Joan, let's see what you've accomplished this week."

I took my usual seat at the desk, and Andrei stood behind me. It felt a little weird with him there. Most people don't think of math as a very personal subject, but I do. It's personal because I put my heart into it, just like Mom puts her heart into her work for Harvest Team and Elena puts her heart into her poetry. One day at the end of junior year Elena tracked me down after school to show me a poem she'd written, a sonnet about the tropical fish in her aquarium. To be honest, the poem was kind of silly, but I came up with something nice to say about it, because I knew how vulnerable she must've felt at that moment. And I felt the same way now as I pulled my math notebook out of my backpack. I could sense Andrei peering over my shoulder.

So I hunched forward a bit, shielding the notebook with my body. I passed it quickly to Professor Taylor, who leaned back in her chair and started looking it over.

This was our weekly routine. Laura would review my work before asking me questions. I was always nervous at this point, worrying if I'd made any mistakes or left something out, and it was hard to distract myself from my worries because the office was so elegantly bare. There was nothing on the big computer screen now except a solid-blue screensaver, and the only decorative item on the desk was a framed photograph of a smiling woman who appeared to be a few years younger than Laura. So all I could do was stare at the woman in the photo and wonder once again who she was.

The picture showed her on a park bench. She had long curly hair and wore a winter coat with a fur collar. She also wore lipstick and pearl earrings, and she smiled at the camera with her eyelids half lowered, as if she were teasing the photographer. The woman was prettier than Laura, but more important, I didn't see any family resemblance. She wasn't Laura's sister. She was probably a girlfriend.

I was kind of thrilled when I saw that photo for the first time, back in September. At the start of my first meeting with Professor Taylor, I pointed at the picture and said, "Oh, who's that? She so pretty." But Laura replied, "Just a friend," and turned back to her computer screen. She changed the subject and talked about number theory for the rest of the hour. When we met again the following week, I was still curious about the photo, but I didn't ask any more questions.

Still, I thought about it every time I went to see her. When I was in Professor Taylor's office, I saw my own future, and I wanted it to include someone like the woman in that photograph.

After about ten minutes Laura scratched the back of her neck, but she kept on studying my notebook. My nervousness got worse—did my equations even make sense? Did I make some stupid error that ruined the whole thing? I heard Andrei breathing behind me, the air whistling through his nose, and I truly regretted bringing him here. I could just imagine how humiliated I'd be if Professor Taylor started criticizing my work in front of him.

Finally, after another ten minutes, Laura looked up from my notebook. "Well, I'm impressed. This is very ambitious." She smiled. "I'm not an expert on string theory, but I know that the subject is notoriously difficult to understand. And yet you seem to have a good grasp of the material."

I smiled back at her. No, I *beamed.* Professor Taylor was a no-nonsense person, so if she said she was impressed, it really meant something. "Yeah, it's crazy hard. But I took classes in topology and differential geometry last year, and that helped me a lot. And almost every research paper on string theory is archived on the Web, so I downloaded a bunch of the papers and read them over the weekend."

"As I said, very ambitious. You've crossed over to a new field of study that I've never tried myself. I'm one of those snobbish mathematicians who steer clear of science. I have an allergy to anything that might be even remotely practical." Laura placed my notebook on her desk and opened it to a page crammed with equations. "I once met a few string theorists at Princeton, including the famous Edward Witten, and I'm familiar with the math they use. But I've never really understood their goals."

I leaned across her desk. I wanted to tell her everything I'd learned, from start to finish. "Okay, it's like this. Right now, there are two basic theories that explain the universe. There's quantum physics, which explains atoms and other really small things, and there's the theory of relativity, which explains really big things like gravity and galaxies. But scientists think there might be an even more basic theory that combines quantum physics with relativity. That's why they're working on string theory, because they think it might lead them to a Theory of Everything."

Laura nodded. "Yes, yes, I knew that much. What I don't understand is why the scientists are still pursuing the idea. How long have they been working on this theory? Since the 1970s?"

I thought of my conversation with Dad, but only for a second. I was too excited to dwell on the obstacles. "Yeah, there's a long history. The theory kept changing, getting more complicated. And some of the physicists gave up on it, because they stopped believing it would ever explain the universe in a useful way."

"So why are you interested in it?"

"Well, I think the scientists are quitting too soon." I reached across Laura's desk and turned the pages of my notebook until I reached the one I wanted to show her. "Here's my plan. According to string theory, there are more dimensions in the universe than the three ordinary dimensions of space—you know, length, width, and height. The theory says there are six extra dimensions that are totally hidden because they're too small to see. At every point in space, the extra dimensions are folded up into teeny-tiny manifolds, like microscopic curls of fabric in a giant carpet. All the physical laws of the universe—the laws of gravity and electromagnetism and atomic forces—depend on the geometry of those tiny crumpled manifolds. And I have a new idea for how the extra dimensions can be folded up."

I leaned over a little more, lifting my butt off the chair, so I could point at the twisted shapes I'd penciled on that page of my notebook. But to my surprise, Professor Taylor didn't look at the figures I'd drawn. Instead, she placed her hand over mine and gave me a sympathetic pat.

"Listen, I know you've worked very hard on this. And your work proves that you've mastered several important branches of mathematics. But hundreds of gifted mathematicians and physicists have already tackled this problem. They've studied it from every angle and published thousands of papers on the subject. So it may be overoptimistic to think that you can accomplish something original in this field."

She kept patting me, her delicate fingers tapping the back of my hand. She'd softened her voice too, making it as gentle as possible, which probably wasn't easy for someone who was so strict and formal. But her words still hurt. I felt dizzy and short of breath. I opened my mouth, but I couldn't speak.

"Joan, the purpose of this tutorial is for you to do independent work on a mathematical problem and hopefully find an original solution. With any luck, you might even be able to publish your solution in a scholarly journal. So your best strategy is to choose a smaller, easier problem to work on, something you can realistically handle."

I pulled my hand away from Laura's and rested it in my lap. Then I took a deep breath. "I thought you said I could work on any problem that I found interesting."

She paused, taking some time to choose her words. "Yes, I said that. And I think it's wonderful that you've discovered a passion for string theory. But I really think you'd be better off if you waited a while before taking on such a difficult problem. A year from now you'll be in college, and you'll already have so many credits in math that you could probably graduate in just two or three years. And then, once you're in grad school, you could find an adviser who specializes in string theory and start working on it in a serious way." She tilted her head, trying to look me in the eye. "In other words, leave the harder problems for the future, when you'll be better equipped to solve them. Does that sound like a reasonable plan?"

It didn't. It was ridiculous. I'd picked a good problem, and I wanted to work on it *right now*. I hated the way Laura was looking at me, her brow creased with concern. It was so condescending, so superior. She didn't think of me as a fellow mathematician. To her, I was just a seventeen-year-old oddity, a weirdly precocious kid who could perform some cute math tricks but couldn't do anything "in a serious way." That had been her attitude all along, but I'd been too dumb to see it before. I'd liked her too much.

I reached for my notebook. I wanted to fling it across the room, but I kept my hands steady and picked it up, glancing at the scribbled diagrams that Professor Taylor hadn't even bothered to look at. I was going to close the notebook and stuff it into my backpack and walk out of her office without another word. It was a stupid, self-destructive decision. I knew it would only confirm Laura's opinion of me as childish and undeserving of adult respect. But I didn't care. I didn't need her good opinion. She was just a math professor, and my allegiance was to mathematics itself.

And to my new idea about string theory, inspired by Samantha.

Before I could close my notebook, though, Andrei stepped forward and bent over the back of my chair. He extended his arm and pointed at the diagram I'd drawn at the top of the page, which looked like a soft pretzel that had been twisted and stretched to the breaking point. "Excuse me, but is that a picture of a manifold? The folded-up extra dimensions you were talking about before?"

I had to crane my neck to look at him. Andrei had pressed his lips together in a long, thin line, giving his face a somber, thoughtful expression. Unlike Professor Taylor, he was taking me seriously.

I nodded. "It's a simplified drawing. The real manifold has six dimensions and all sorts of twists and turns, so it's impossible to draw it accurately on a two-dimensional sheet of paper. That sketch was the best I could do."

Andrei found this amusing for some reason, and he let out one of his squeaky laughs. "You know, I studied this type of manifold in my topology class at Lyceum 239. It's called a Calabi-Yau space, correct? Named after two great mathematicians, Eugenio Calabi and Shing-Tung Yau?"

I nodded again. "That's right. Calabi conjectured that this kind of manifold could exist, and Yau proved its existence. Then some physicists discovered that they could use these manifolds to describe the shape of the extra dimensions in string theory. But they ran into trouble because there were jillions of ways to fold up the dimensions. That gave them a really huge number of manifolds to choose from, and the physicists had no idea which was the correct choice, the manifold that actually reveals the true shape of the universe."

I looked at Professor Taylor out of the corner of my eye. She was frowning. We'd left her out of the conversation, and she clearly didn't like it.

Andrei extended his arm a little farther. The tip of his index finger touched the left side of my diagram, where the pretzel-like manifold was stretched the most. "It looks like you've deformed this part of the Calabi-Yau space for some reason."

The boy was definitely a quick study. I nodded a third time. "Yeah, I'm playing around with it. I'm experimenting with different changes to the manifold, warping and twisting and mangling the thing. I'm hoping that I get lucky and create a new version of the manifold that seems more natural than all the others. And maybe that way I'll solve the problem and find the true shape of the extra dimensions."

Andrei thought it over for a few seconds. He raised his hand and rubbed his chin, caressing the trio of pimples there. Then a big dopey grin spread across his face, the smile of a crazed Russian clown. "Joan, this is wonderful! I want to work on this problem with you. Would you be willing to consider a collaboration?"

He leaned over me, breathing fast. I wrinkled my nose, smelling that mix of cabbage and mothballs again, but I was also pleased at the same time. I smiled back at Andrei, then turned to Professor Taylor. "What if both of us worked on it? Would that be more acceptable?"

She stared at us, still frowning.

Andrei jumped in before she could answer. "Oh, yes, Professor Taylor, please say you'll agree! It's true that we are young, we are beginners, but with both of us working together we'll be able to apply twice as much brainpower to the problem. Our chances of success will be twice as great!"

Laura shook her head. The creases in her brow deepened. "I don't think you've been listening, Mr. Mishkin. *Hundreds* of brilliant people have tried to solve this problem and failed. Your chances of success are minuscule."

Andrei shrugged. "Then I suppose we'll fail. But failure can be good too."

"I'm sorry, I don't—"

"I've failed so many times before with so many different problems, I can't even tell you how many times. But I'm still here, yes?"

She narrowed her eyes, glaring at him. Laura was obviously angry and embarrassed, because Andrei had just demolished her objections with the perfect counterargument. Failure always lays the groundwork for success, in mathematics and everything else. That was a fundamental truth, an axiom, as logically self-evident as the statement that x equals x.

After a few more seconds, Laura turned away from Andrei and looked at me. "You can do whatever you want, Joan. But your grade for this tutorial will be significantly lower if you show poor judgment in choosing your problem."

I snorted. I couldn't help it. Did she really think I cared about my *grades*? "I'm fine with that. I'll take the risk."

"All right then." She glanced at her watch, a fancy silver thing that dangled from her wrist. "Well, our time's almost up. Please shut the door on your way out."

She reached for her computer's mouse and clicked the button on it. The big screen on her desk came to life, displaying half a dozen windows, and she started reading and deleting her emails. I checked my own watch and saw that we still had fifteen minutes left in our hour, but Laura was done. She'd given up on me, and I was fine with that too.

I stood up and put the notebook in my backpack. Andrei bowed again, very low this time. "Thank you, professor, a trillion thank-you's! We'll make you very proud of us!" Then he flashed me a triumphant look, his eyes shining with delight.

And I felt triumphant too, at least for a moment. But just before I left the office, I looked over my shoulder at Laura and wondered if I'd made a mistake. She wasn't a bad person. She was just trying to guide my mathematical education and protect me from disappointment. So why did I get so mad at her? Why did I reject her advice so completely?

I *had* made a mistake. But I wouldn't find out how bad it was until much later.

Chapter Eight

After leaving the Math department building, Andrei and I went our separate ways. He had a gym class at 1:45 with Mr. Plotkin—a loud, hairy, whistle-blowing idiot—and afterwards he had to go downtown to help out his mother, who knew very little English and was having trouble getting telephone service for their new apartment on the Lower East Side. But we made plans to meet the next day at three o'clock to talk some more about string theory.

In the meantime, I headed back to Franklin for a meeting with Ms. Mosner, the high school's very annoying guidance counselor. I really hated those appointments. Mosner spent most of the time struggling with her computer, trying to find my grades and test scores, and once she located my information she always scolded me about my attitude toward college applications. She said I should apply to more schools and visit more campuses and send fawning emails to every admissions officer in the Ivy League. It was a horrible process, and Ms. Mosner made it worse. She was short and ferrety, with small brown eyes and a sharp pink nose, and she seemed to feed on the students' fears and anxieties.

But the worst part of that day's appointment came at the very end, while Mosner was criticizing the essay I'd written for my applications. The essay was about math, of course—that was the subject I wanted to spend my whole life exploring—but Mosner thought it sounded "a little dry." She suggested that I write about the cross-country team instead, and I told her I wasn't interested. Then she leaned forward in her chair and lowered her voice and started talking about my sister.

"Joan, I know what you've been going through the past few months. I lost my own sister to cancer three years ago, and it was a terrible thing. And it's probably worse for you, because the story was in all the newspapers. Now, I don't know if you're ready to write about it yet, but if you are, it might make a really powerful essay. Colleges love to hear about students who overcome adversity."

I stood up and left her office. There was no other sane response.

* * *

I spent the next six hours at City College's main library. Now that the cross-country season was over, my afternoons were free, and I planned to devote all of them to Calabi-Yau manifolds.

The library's collection of math and physics books was excellent. I found a good textbook on string theory and started reading about the physicists who invented it. During the 1970s they began to question the most basic assumption of science: that the smallest things in the universe were particles. Instead, the theorists proposed the existence of microscopic strings that vibrated within the curled-up manifolds of extra dimensions. In this scheme, the strings were the building blocks for every object in the universe. All the fundamental particles—electrons, muons, neutrinos, whatever—were really strings vibrating in different ways; an electron was a string jiggling in one kind of pattern, and a neutrino was a string doing a different kind of dance. Beams of light were made of strings too, and so was the force of gravity. It was a theory that could explain *everything*, but only if the mathematics worked out.

And the math turned out to be crazy complicated. There were actually five versions of string theory: Type I, Type IIA, Type IIB, SO (32), and $E_8 \times E_8$. Although the physicists assumed that all five versions were part of a bigger framework called M-theory, they had no clue what the equations of this theory were. (Edward Witten, the discoverer of M-theory, said the "M" could stand for either "Magic" or "Mystery.") But the toughest problem was dealing with the manifolds, because there were simply too many kinds. At first, physicists worried that there might be millions of ways to fold up the extra dimensions into manifolds, but that was a serious undercount. They later boosted the estimate of possible Calabi-Yau spaces to $10^{272,000}$. I'll give you an idea how big that number is: if you wrote it out in decimal notation—that is, 1 followed by 272,000 zeroes—it would fill up a hundred-page book.

It was a freakin' huge problem. Because all the manifolds had different shapes, the strings inside them would vibrate in different patterns, producing different types of particles and forces. So if you wanted to formulate a string theory that described our universe—a Theory of Everything that would correctly predict the properties of all the particles and the strengths of all the forces—you needed to choose the correct manifold first, selecting it from among the $10^{272,000}$ possibilities. It was like finding a needle in an infinite haystack.

After a while I closed the textbook and started reading another, but the news didn't get any better. I pulled out my math notebook and started drawing Calabi-Yau spaces, trying to deform their shapes in interesting ways, but I knew that making random guesses was hopeless. What I needed was a guideline, a logical principle that would help me construct the correct manifold. I was starting to realize that maybe Professor Taylor had been right. Maybe solving this problem was beyond my abilities.

Then I remembered Dad standing in the middle of my bedroom, his face bright red, full of rage and despair. *What's the point of life if there's no explanation? If the reason for the universe is hidden forever?*

At just that moment, strangely enough, I got a text from Dad. My iPhone buzzed and I looked at the screen.

Where R U?

It was nine o'clock—not very late, but my parents had become more anxious about my safety after what happened to Samantha. Although I didn't like the extra scrutiny, I was willing to humor them, at least until I went off to college.

Studying at library

I stared at the phone for several seconds, waiting for Dad to reply. It took him forever to tap out even the simplest text.

Come home for dinner! Mom made burgrs!!!!!!

He was worried. I could tell from the number of exclamation points. I would've put up more resistance if I'd been making progress on my work, but I was going nowhere. And besides, I was starving.

OK going home now

I shoved my notebook into my backpack and shelved the textbooks. Then I left the library and hurried toward the subway station. The sun had set a couple of hours ago and a full moon had risen above the buildings to the east, shining on the parked cars and sidewalks of Broadway.

It was way past rush hour, so the station was almost empty. At one end of the platform were three tall girls in City College sweatshirts—they looked like they played for the school's basketball team—and at the other end was a homeless woman in a sooty brown coat, sitting on a bench. I walked away from the basketball players because they were talking too loud, and I headed for the homeless woman, who'd parked an overstuffed shopping cart beside her bench. My timing was lucky: in less than thirty seconds, the downtown 1 train roared into the station. I quick-stepped toward the subway car at the front of the train and waited for the doors to open. I'd be home in twenty minutes. I could already taste the hamburgers.

"Hey, dearie? Come over here and help me with this."

The homeless woman had stood up and gripped the handle of her shopping cart, but she couldn't move it. Inside the cart were a pair of suitcases, a grayish blanket, a plastic bag bulging with empty cans and bottles, and a portable television set with a cracked screen. I could see why the woman was having trouble pulling the cart—she was less than five feet tall and at least seventy years old. Her face was the color of cardboard and laced with hundreds of wrinkles.

The subway doors opened as I stepped toward her. "You want to get on the train?"

She frowned. The wrinkles around her mouth multiplied. "What do you think? Of course I want to get on! Help me pull this thing!"

Her voice echoed against the station's walls. Startled, I grabbed the cart's handle and gave it a tug. The wheels whined in protest and the cart yawed across the platform.

"Hurry!" The old woman let go of the handle and toddled into the subway car. She stood just inside the doorway and waved me toward her. "Come on, come on!"

The train's loudspeakers let out their warning signal—*dee-doo!*—and the subway doors began to close. I rushed backward into the train car, but the doors slammed into both sides of the cart, rattling the heck out of it.

"Careful!" The woman scowled and pointed at the cart. "Watch what you're doing!"

The doors bounced off the cart and opened wide enough for me to yank it into the train. The woman hissed a sharp tsk of disapproval, then lowered her butt onto a nearby seat. Plenty of seats were available, because the train car was empty except for us. "Bring it over here, dearie. I need to see if you broke anything."

I was seriously annoyed by now, but I dragged the cart next to her seat. A moment later, the subway doors closed all the way and the train jerked forward. It crept out of the station, moving much more slowly than before, probably because there were red signals on the track ahead.

I sat down across from the woman while she inspected the contents of her cart. She perched on the edge of her seat and bent over so she could examine the suitcases and the television set. There were pink sties on her eyelids and thin black hairs on her upper lip. The hair on her head was gray and fastened with a rubber band, and under her coat she wore a threadbare beige sweater that stretched tightly over her breasts and stomach.

She reminded me of Grandma Rose, my dad's mother, who'd died when I was eleven. I'd never liked her—she was sour and snappish, a lot like this old woman on the train. But I told myself to calm down and have a little more sympathy. Being homeless in Manhattan was the ultimate torture. You had to suffer in plain sight of the richest people in the world and watch them do nothing.

After half a minute she finished her inspection and looked up at me. "You're lucky. Nothing's broken. But that was a close call, eh?"

I put on a smile for her and nodded. "Where are you going? Do you have someplace to stay tonight?"

She looked to her left and right, taking in the empty subway car. "What's wrong with right here?"

"Nothing, I guess. But my mom works with a lot of homeless shelters, and she says the ones for women are actually pretty safe. The best intake center is in the Bronx. You can get there on the uptown 2 train, if you want."

The woman laughed. It sounded like the rattling of her cart. "I'm not homeless. The whole world is my home." She leaned forward in her seat. "Don't you know who I am, Joan?"

Her voice had dropped to a low, confiding tone, as if we were old friends. But it didn't make me feel at ease. It had the opposite effect. "Uh, how do you know my name?"

She shrugged. "Oh, I know everything. And we already met, just two days ago. Remember? On the trail at Van Cortlandt Park?"

I shook my head. Not because I didn't remember, but because I didn't want to. "I'm sorry, I don't know what you're—"

"Come on, don't play games." She frowned again. The wrinkles fanned out from the corners of her eyes. "You've been thinking about it ever since. Wondering why no one else could see the boy on the trail, and why he wore that number on his shirt. And worrying that you were going crazy."

I rose to my feet, slow and unsteady. I felt like I was going to tip over, even though the train was barely moving. "Uh, no. That's not—"

"Here's the good news: You're sane. You're not having a psychotic breakdown. I'm as real as anything else here." The woman slapped her hands against her coat, and gray swirls of dust rose from the fabric. "See?"

I grabbed the subway pole to my left and held it tight. "Okay, listen, I need to go."

"You're confused because I look different from the last time you saw me. But I can take any shape I like—a boy, an old woman, an animal, whatever. My favorite shape is fire, actually. You've heard the story of the burning bush?"

I spun away from her. Because the subway car was at the front of the train, the only exit was the door at the car's back end. The sign on the door said, "Riding or moving between cars is prohibited," but at that point I didn't care about the rules. I was going to open the door and dash into the next car and get as far away from the homeless woman as I could.

But the door was locked. The handle wouldn't turn.

Which wasn't so unusual. Sometimes the train conductors locked the doors between the cars. I knew I shouldn't get freaked about it.

And yet a cold dread settled in my stomach. I turned around and looked at the old woman, who was still frowning at me. Her eyes narrowed and her face seemed to darken, becoming sharp and awful, like something out of a nightmare. "You need to calm down, Joan. I knew this encounter would be hard for you, so I gave you a couple of days to prepare yourself. But we can't wait any longer. Time is growing short."

As I stared at her, the train rumbled out of the subway tunnel and onto the elevated tracks that run above Broadway between 120th and 130th streets. Through the train car's windows I saw the apartment buildings and warehouses on the western edge of Manhattan, lit by the full moon and a thousand streetlights. Then I got a glimpse of the next subway stop, the platform of the 125th Street station, and I felt a burst of hope.

I can escape! I'll bolt out of the train as soon as it pulls into the station! Then I'll hustle downstairs to the street and sprint down Broadway, and I won't stop till I reach my building on 78th Street!

The old woman shook her head. She looked disappointed. "You're not listening. That's too bad. Now I have to take extraordinary measures."

She raised her right hand and snapped her withered fingers. An instant later, the train screeched to a halt.

The subway car stopped a hundred feet short of the 125th Street station. Then the conductor's voice came over the car's loudspeakers: "*Attention, passengers. We're being held here by the train's dispatcher. We should be moving shortly.*"

The woman lowered her hand. “I hate doing that. The delays on this subway line are bad enough already.” She patted the seat next to her. “Sit down, Joan.”

I stepped toward her. I couldn’t escape this craziness. I was trapped in the train car with a phantasmal woman who seemed to be reading my mind and controlling the New York subway system. The only reasonable explanation was that I was hallucinating. I was imagining the existence of a cranky old woman with Godlike powers. And the only way to wake up from this dream was to fight it. I needed to confront this messed-up figment of my imagination and drive it out of my head.

Instead of sitting next to the old woman, I stood in front of her and looked her in the eye. “What do you want from me? What do I have to do to make you go away?”

She returned my stare, unflinching. “You’re already doing it, believe it or not. You’re on the right path. The challenge will be staying on it.”

“Yeah, you know what? I have no idea what that means.” I bent over and got in her face, leaning forward until my nose was just inches from hers. “Why don’t you leave me alone, okay?”

“Because you need my help.” She folded her arms across her chest. Her eyes were wet and gleaming. “I’ve given you all the tools you’ll need for this task, but I’m not sure you’ll have the strength to carry it out.”

This was frustrating. My hallucination was speaking in riddles. “You’re not making sense. What ‘tools’ are you talking about?”

She smiled. Her face turned almost pretty, even with all the sties and wrinkles. “What are you good at, Joan? What are you most proud of?”

“Math? You mean mathematical tools?”

She nodded. “Exactly. That was my gift to you.” She unfolded her arms and pointed at my forehead. “I gave you a very special skill. No one else has ever had it, not in all of human history, thousands of years. It’s a beautiful thing, and it’s all yours.”

Her voice became softer, less snappish. She wasn’t like my sour Grandma Rose anymore; now she was more like the grandmother I’d always wanted, a wise, kind woman who would’ve loved and protected me. In other words, the illusion had transformed into something more comforting, a maternal figure who didn’t frighten me so much. But it was still an illusion. I still wanted it out of my head.

“What about the ‘task’? What’s that all about?”

“You already know. It’s solving the problem you chose. The one you’re working on in your notebook.” She pointed at my backpack. “By the way, I liked how you stood up to your professor this afternoon. That took a lot of courage. And I was glad to see you make friends with the Russian boy. He can help you."

I stood up straight and stepped back. "String theory? That's the task you're giving me?"

"You gave it to yourself, really. But like I said, you'll need some help." She stopped smiling and leaned forward. "You see, this is an important moment in history. The human race is finally ready to learn the details of my plans, how the universe began and how it really works. You're going to solve the equations of string theory and reveal the true nature of the cosmos. And by revealing my mathematical secrets, you're going to end this Age of Ignorance and set humanity on a better course. Wisdom will lead to peace, and enlightenment will bring happiness. But things could go wrong if it's not handled carefully."

She shifted in her seat, squirming a bit, as if she were uncomfortable. She clearly knew the next question I was going to ask.

"Things could go wrong? What does that mean?"

Grimacing, she scratched her cheek. Her fingernails rasped against the loose skin. "It's hard to explain. There are crucial times in history when the world plunges into chaos. The old order falls apart and everything gets rearranged. At those times, I can't control what happens." She moved her hand a little lower and scratched her chin. "The universe is so complex, I can't dictate its future. There are other forces at work, fundamental impulses that have been evolving for billions of years, and sometimes they oppose me. They can alter my plans and destroy what I've created."

I was frustrated again. My hallucination wasn't playing by the rules. This phantasmal woman was obviously supposed to be God, and that meant she had to be all-knowing and all-powerful. So why was she talking about things she couldn't control? The illusion was illogical. I couldn't even fantasize properly.

The old woman looked tired. She let out a sigh. "You know what your problem is? Like everyone else, you grew up with certain assumptions about God and the universe. And they're all wrong. That's why you're so confused right now."

I shook my head. I'd had enough. I'd tried reasoning with my phantasm and asking her questions, and it hadn't driven her away. My mind was still torturing me with this stubbornly solid illusion, this homeless woman who wouldn't leave me alone. So now it was time to get serious.

I took a deep breath. I was going to scream.

"*That's enough! Go away! Get out of my head!*"

My voice echoed against every surface of the subway car—the steel walls, the hard seats, the closed doors, the scuffed floor. It was loud enough to make the handrails vibrate and the windows shake in their frames. But it had no effect whatsoever on the old woman. She just rolled her eyes.

"You're not thinking straight, Joan. You can't be hallucinating, because I'm telling you things that you never could've imagined on your own." She pointed at me. "For instance, I can tell you something about your sister. She thought about you at the end. You were the last person she pictured just before she died."

I clamped my hands over my ears and screamed again. I was trying to be as loud and angry as possible. I thought if I could ratchet up my fury and reach a high enough level of righteous outrage, I could exorcise the old woman and hurl her out of my mind.

"*GO! JUST GO! I DON'T WANT YOU! I DON'T BELIEVE IN YOU!*"

This time, my screams had an effect, but unfortunately it wasn't what I'd intended. The old woman stood up and grabbed my arm just above the elbow. With surprising strength, she pulled me toward one of the windows on the left side of the subway car.

I was stunned and scared and tried to wriggle out of her grasp, but she held me tight. She stood next to me and leaned sideways, bringing her face close to mine. I could feel her hot breath on my cheek. "Look out the window. You see the street? The people?"

I was too frightened to disobey her. I gazed out the subway car's window, looking down from the elevated train tracks at the traffic on 125th Street. Dozens of cars and taxis streamed in irregular lines down the street, jockeying for position. Their headlights shone on the people walking along the sidewalks. Above them all, the full moon cast its glow, painting the city silver-blue.

I nodded. I was shaking from head to toe. "Yes, yes, I see. Now let me—"

"You'll believe I'm real once you witness this. I did the same thing for Joshua when he was fighting the Amorites."

"What? Amorites? What are you—"

"*Look!*" She tightened her grip on my arm, and at the same time she pointed at the full moon in the eastern sky. "*Moon, stand thou still over New York!*"

The heavens seemed to blink. There was no crash, no explosion, no noise at all. But in that instant, everything stopped moving.

The cars and taxis halted on 125th Street. It looked like all their engines had died at once, but the vehicles didn't roll or slide to a stop. They simply froze in place. The people on the street froze too, caught in mid-stride. Each man and woman stood there like a mannequin, carefully posed, with one foot planted on the sidewalk and the other hovering a few inches above. It was as if the city had suddenly become an enormous diorama, an unbelievably detailed exhibit in a gargantuan museum, intended to show a single moment in the history of New York. It was astonishing, mind-boggling.

And horrifying.

I grew dizzy. My guts churned and my throat tightened. The stillness and silence felt like a weight on my chest, so crushing that I had to gasp for breath. But the worst part was the lights. The headlights and streetlights of the city no longer cast their beams in straight lines. The light rays curved in crazy patterns around each bulb, forming twisted, misshapen halos. And the brightest halo of all surrounded the full moon, its warped light blazing like frozen fire.

I slumped against the old woman. My vision was darkening. I was losing consciousness. "Please…enough…the lights…*the lights!*"

She held on to my arm. It was the only thing that kept me from collapsing. "Yes, Joan, I know. That's what happens to a beam of light when time stands still."

Then I lost all feeling and everything went black.

Chapter Nine

The next thing I saw was Mom's face.

I lay on a gurney and Mom sat beside it, pale and frantic, gripping the steel bedrail with both hands. Dad stood behind her, bathed in the fluorescent light of the room, which was bare and windowless and much too bright. He was staring at the screen of some machine beeping next to the gurney—a heart-rate monitor?—but Mom's eyes were fixed on me. As soon as I woke up, she jumped out of her chair and leaned over the bedrail and clutched the short sleeves of my hospital gown.

"*Joan!* Oh thank God, you're awake!"

I looked at her carefully, examining her face as if I'd never seen it before. There were faint black smudges running down the sides of her nose. Her tears had smeared the black liner on her lower eyelids. My own eyes watered at the sight of it.

"Hi, Mom." My throat felt like sandpaper. I could barely talk.

Dad rushed over to the other side of the gurney. I could tell from his pinkish cheeks that he'd been crying too. "Are you okay, Joanie? How do you feel?"

I flexed my arms and legs under the gurney's blanket, checking to see if anything was broken. There was an IV tube attached to the back of my right hand, and I had a sore throat and a bit of a headache, but otherwise nothing seemed to be wrong. And yet here I was in the hospital, and I had the feeling that several hours had passed since I'd blacked out. So I thought back to the last things I could remember—the ride on the downtown 1 train and the homeless woman with the shopping cart and the seriously disturbing things she'd said. Then I remembered the very last thing, how the traffic on 125th Street stopped and the whole city froze.

The horror came back in full force, wringing my body. My stomach heaved, but luckily it was empty. The only thing that came up was a drop of acid that seared the back of my throat.

Mom was watching me like a hawk, so she noticed my distress. "What is it? What's wrong? Talk to me, Joan!"

I shook my head. How could I explain it? I couldn't even try. "No, I'm okay. What hospital is this? I don't remember coming here."

Mom didn't answer. She bit her lower lip and turned to Dad, which meant she was afraid to say something.

Dad bent over the gurney and patted my shoulder. "The paramedics picked you up from the subway station on 125th Street. You were staggering back and forth along the platform, and they were afraid you might fall on the tracks." He pointed at something on the floor that I couldn't see. "Luckily, they found your backpack and all your contact information, and they called us right away. That was four hours ago."

I didn't understand. I'd lost consciousness while I was still in the subway car. I had no memory of getting off the train. "No, I was on the subway, heading home. I don't remember going on the platform. Or seeing any paramedics."

Mom reached over the bedrail and patted my other shoulder. It was like a contest between her and Dad, to see who could comfort me the best. "It's not your fault, Joan. We know you didn't do anything wrong. The doctors tested you for drugs, and everything came up clean."

I winced. Drugs would be everyone's first assumption, of course. "So is this Saint Luke's?" That was the hospital closest to the 125th Street station. "How long do I have to stay here?"

Mom turned again to Dad. He bent a little lower and squeezed my shoulder. "Actually, this is Bellevue. You know, the hospital on First Avenue."

I knew about Bellevue. Anyone who watched TV was familiar with the hospital. It was the place on all the crime shows where the New York police brought the lunatics they found on the street. "Bellevue? I'm in a psychiatric ward?"

Mom shook her head fiercely. "No, this is part of their emergency room. We're not gonna let them admit you to the hospital. We're gonna get you out of here and bring you home."

"I don't…why would they admit me?" I lifted my head off the pillow and turned to the left and right, surveying the bare room. I focused in particular on the door, which had a square of reinforced glass at eye level. "What did I do?"

Dad squeezed my shoulder again, but now his grip felt weaker, less reassuring. "When the paramedics found you on the subway platform, you were in a hysterical state. Ranting and raving, they said. They thought you might be a danger to yourself, so they brought you here."

My headache sharpened. I was in big trouble now. The doctors wanted to commit me to the psychiatric center. And they were *right*—I *did* belong in the hospital, because I *was* having hallucinations. I'd imagined I'd seen God, and that was a classic symptom of craziness. But the scariest part was that I couldn't say for sure what had actually happened in the subway car. Had I really hallucinated the homeless woman? She'd seemed so real.

To be honest, I didn't know what to think. I was terrified.

I grabbed Mom's hand. It was instinctive, a baby reaching for her mother, like I'd gone back in time and become a helpless infant. "Don't let them commit me! I want to go home!"

She twined her fingers with mine. At the same time, she raised her other hand to my head and brushed back my hair. "Shhh, don't worry. Now that you're awake, we'll get the doctors to release you. We'd never let you stay in a place like this."

Dad nodded. "They've already done a CAT scan on you, and they couldn't find anything wrong. So now I just need to track down the emergency-room doctor and get him to sign the release papers." He stepped backward, edging toward the door. "Sit tight, okay? I'll be right back."

He opened the door, letting in a sudden blast of noise from the corridor. Someone was screaming "*JESUS! JESUS SAVE ME!*" at the top of his lungs. A second later, a team of doctors and nurses in green scrubs barreled past the doorway, pushing a gurney that held the screaming man. Two of the doctors gripped the man's arms, pinning him to the gurney as they raced down the hall, but I got a fleeting glimpse of the man's face. It was gaunt and unshaven and bleeding from cuts on his forehead and chin.

"*THIS IS HELL! SAVE ME FROM HELL, JESUS!*"

Dad stood there in the doorway, frozen, staring at the gurney as it sped past. Then Mom shouted, "Shut the door, Joe!" and he snapped out of his trance. He closed the door behind him and went off to find the emergency-room doctor.

Strangely enough, I didn't freak out at the sight of the screaming man. Instead, I took a deep breath and steeled myself. All at once, I was full of outrage and determination. I didn't belong in the same place as that man. If the doctors thought I did, then they were the crazy ones.

I let go of Mom's hand, threw the blanket off me, and sat upright on the gurney. "What should I tell the doctor? So I can convince him to let me out of here?"

She gave me a serious look, lowering her eyebrows. "You didn't get much sleep last night, did you? I woke up at 4 a.m. to go to the bathroom and I noticed that the light was still on in your bedroom."

"Yeah, I was working on a new math problem. I slept for only a couple of hours."

"Mention that to the doctor. Tell him you were lightheaded when you got on the subway." She cocked her head and thought it over for a second. "Plus, it was after 9 p.m. and you hadn't eaten dinner. The combination of those things could make anyone go a little haywire."

I wasn't sure that would be enough. I took another deep breath and tried to think of a good excuse for my breakdown. But before I could say anything, the door opened and Dad came back inside with the doctor.

He was a thin, middle-aged guy dressed in scrubs, very pale and nearly bald. His eyes were bloodshot and the fringe of hair remaining on his scalp was stark white. He looked like an unhappy vampire, his lips pressed together to hide his fangs. He glanced for a moment at the clipboard in his hands, then approached my gurney.

"Hello, Joan. I'm Doctor Harrison, one of the emergency-room psychiatrists here. Your parents tell me that you want to go home?"

I nodded vigorously. "Yeah, I'm feeling a lot better."

"Well, let's take a look at you." He turned on a penlight and aimed its beam at my eyes. "I'm going to ask you a few questions to see how alert you are, all right? What's the name of the current president?"

"Ugh, don't remind me." I made a joke to prove how sane I was. "I'd rather not say the jerk's name. But I can tell you the name of the woman who should've been elected. It rhymes with Millary Minton."

Mom let out a chuckle and covered her mouth to stifle it. But the doctor frowned. "Better to avoid politics here. It just makes people upset. Can you count backwards from one hundred for me?"

"Oh, I can do better than that. I'll give you the list of the first ten Eisenstein primes: 2, 5, 11, 17, 23, 29, 41, 47, 53, and 59. They're the prime numbers that can be written as three times *n*, minus one, where *n* is either one or an even integer."

Doctor Harrison didn't seem impressed. He lowered his penlight. "Yes, your parents mentioned that you love math."

I nodded again. "That's why I stayed up so late last night, because I was working on a new problem. So I was exhausted all day, and I guess it finally got to me."

"Has this ever happened to you before? Getting excited about a math problem and working on it till you're exhausted?"

I sensed a trap. The doctor wanted me to admit that I had some kind of disorder. "No, no, it was very unusual. I usually get lots of sleep."

I glanced at my parents, who knew this was a lie, but neither said anything. They wanted me out of Bellevue, so they weren't going to contradict me. Harrison glanced at them too, then studied his clipboard.

"What about religion? Are you a religious person?"

I felt a jolt of adrenalin. Why was he asking about religion? How much did he know? "Uh, no, not really. I mean, Mom is Catholic, and Dad is half-Jewish, half-Unitarian. So I don't know what that makes me. Maybe a Catho-Jewnitarian?"

Harrison tapped one of the papers attached to his clipboard. "The paramedics reported that when they found you on the subway platform, you were shouting religious statements." He held the clipboard close to his nose and squinted. "You yelled 'God is fire' dozens of times, they said. And after they calmed you down, you started muttering 'Joshua and the Amorites' over and over. That's from the Bible, isn't it?"

I wanted to jump off the gurney and dash out of the room, but I managed to control myself. I shrugged, raising my hands in the air. I hoped it didn't look too fake. "I don't know. I don't remember saying any of that stuff. I must've been babbling nonsense because I was so tired."

Harrison looked up from his papers. "I have to be honest with you, Joan. This behavior is worrisome. Your parents gave me some background information and told me about the death of your sister a few months ago. I suspect you've experienced a lot of emotional turmoil since then, and the strain may have triggered this episode."

I shook my head and grabbed the bedrails. This wasn't going well. "Look, I'm fine now, see? I just want to go home and get some rest."

"I understand that, but I don't think you should leave just yet. I'd like to keep you under observation here for 24 hours. Then we can do some more tests and make sure you're okay."

While I sat there in shock, Dad stepped toward the gurney. He stood next to the bedrail, putting himself between me and Harrison. "Doctor, we want to take her home now. This isn't the right place for her."

Harrison leaned closer to Dad and lowered his voice. "Mr. Cooper, can I speak to you in private for a moment? I think—"

"No, I've heard enough." He raised his hand and chopped the air. "She's obviously not a danger to herself or anyone else, so you have no legal right to keep her. So please sign the release form and we'll be on our way."

Mom stepped forward too. She positioned herself at the foot of the gurney, outflanking the doctor. "We'll get a therapist for Joan somewhere else. We live on the West Side, and there's no shortage of psychiatrists up there. I don't think we'll have any trouble finding a good one."

Harrison grimaced. I got the feeling he wasn't accustomed to dealing with middle-class parents. Most of the psychiatric patients at Bellevue were poor and homeless, so I guess it was easier for him to boss them around. "Your daughter's condition may be more serious than you realize. I haven't had a chance to do a proper diagnosis yet, but Joan may have suffered a schizophrenic break tonight. And if that's the case, it will definitely happen again, and it'll be much worse next time."

The word "schizophrenic" startled Mom and Dad. Both of them stepped backward when they heard it. A look of grim satisfaction appeared on Harrison's face when he saw their alarm. At that moment I hated him more than I'd ever hated *anyone*.

I leaned forward on the gurney and pointed at him. "You're a lousy doctor, you know that? You just said you haven't had a chance to diagnose me, and that means you don't really know what's wrong. So instead of scaring my parents, you should keep your mouth shut."

That wiped the satisfied look off his face. He glared at me, then pulled a pen from the pocket of his scrubs and signed one of the forms on his clipboard. "All right, I tried my best. My job as a psychiatrist is to warn patients of possible dangers. But if you choose to ignore those warnings, there's nothing I can do."

"I'm not ignoring you. I'm listening very carefully." I glared back at him. "I just want a second opinion. And I feel sorry for all the other patients here who can't afford to get one."

Harrison detached the signed form from the clipboard. "There's something else you should know before I release you. The paramedics said there was a crowd of people watching you at the subway station, and some of them were using their phones to take videos of your psychotic break." He stepped toward Dad and thrust the release form into his hands. "So if you want to learn more about the episode, you can probably go on YouTube and watch the whole thing."

Then he turned around and marched out of the room.

Chapter Ten

I spent the next twelve hours in my bedroom. My parents said I could stay home from school, so I slept until noon, which was the easiest way to avoid all my problems. After I woke up, I stayed in bed for another hour, trying to go back to sleep and failing miserably.

At 1 p.m. I finally gave up and reached for my laptop. Within two minutes I found the YouTube video.

It was easy to track down—I just put "crazy subway" in the website's search box, and the video appeared at the top of the list. Someone named SirBob1429 had uploaded it to YouTube under the title, "Crazy God Girl in NYC Subway Station." I recognized it right away because the picture next to the title showed a close-up of my T-shirt, the one with the slogan **I AM WOMAN**. In the sixteen hours since the video had been uploaded, it had been viewed more than five thousand times.

I didn't want to watch it. I was afraid I'd see myself do something unforgettably humiliating. But five thousand people had already seen the video, and there was a small but not insignificant chance that at least one of them was a fellow student at my high school. And if one person from Franklin High saw it, the entire school would know about it by tomorrow morning. I had to be ready for that.

Cringing, I clicked on the video.

A moment later I saw myself on the laptop's screen. I was on the outdoor train platform at 125th Street, the station on the elevated tracks that ran above Broadway. I stood pretty close to the platform's edge, which jutted several feet above the moonlit tracks. Behind me were the night sky and a horizon crenellated with distant buildings, and in front of me were two paramedics wearing dark blue uniforms and light blue gloves. The video was shaky—SirBob1429 was laughing as he aimed his phone's camera at me. But he stood only a few feet behind the paramedics, so he got a pretty good view of my face.

Although I was retreating from the men in uniform, I didn't look scared. My face glowed under the station's harsh lights, and my eyes glittered with manic intensity. I pointed my right hand at the paramedics and smiled at them, wide-eyed, as if I'd never seen or imagined such wonderful public servants. My pimples were all-too-visible on my forehead, and my hair was a disaster, a tangled black mess, but all in all, I didn't look as terrible as I'd feared. I seemed excited and happy, like a little girl at her birthday party. I spoke in a wild, enthusiastic torrent, my voice straining to keep up with all the things I had to say.

"God is fire! God is energy, God is physics, God is *transformation*! God can take any shape She likes—a boy, an old woman, an animal, whatever! God is the principle, the logic, the guideline, the theory!" I raised both hands above my head, open-mouthed and ecstatic. "All our assumptions about God and the universe are wrong! God is the fundamental impulse! Our great task is to understand the cosmos, and our best tool is mathematics!"

I paused the video. I needed a few seconds to catch my breath. There were so many disturbing things about this scene, but the first and most obvious problem was that I didn't remember *any of it*. I recognized some of the things I was saying—I was repeating the nonsense I'd hallucinated in the subway car—but I had no memory of shouting them at the paramedics. Even worse, this girl in the video was nothing like me. She had no doubts, no inhibitions. She was crazy.

I closed my eyes. I still had the headache that had started bothering me at the hospital. I raised my hands to my temples and rubbed them in slow circles until the pain eased. Then I opened my eyes and restarted the video.

On the subway platform, I dropped my arms to my sides as I finished my rant about God. SirBob1429 shouted, "You tell 'em, girl!" from behind his phone, then turned the camera away from me and focused on the small crowd of late-night commuters who were watching the scene. Some laughed, some shook their heads. Some cursed. New Yorkers are usually pretty tolerant of crazy people, but I was disrupting train service at the station, and that was never a good thing.

After a moment, the camera turned back to me and the paramedics, who were now moving toward the edge of the platform. They were stepping between me and the edge, probably to make sure I didn't throw myself onto the train tracks. The taller of the two men held out his gloved hands, palms up, in the universal gesture of good will. "Hey, let's talk, all right? We're here to help. My name is Dave. What's yours?"

I pointed at him, clearly delighted by what he'd just said. "You see? *You see?* You're a marvel, a miracle! You have a gift, and no else has ever had it, not in all of human history! The Age of Ignorance is ending, and soon we'll discover how the universe began! We'll know the mind of God, because God is fire!"

The shorter paramedic didn't say anything, but he slowly moved sideways across the platform. Clearly, his job was to circle behind me while his partner held my attention. Dave the taller paramedic rested his hands on his hips, calm and casual. "All right, that's very interesting. I like talking about God. But why don't you tell me your name first? Do you live around here?"

"I live everywhere! The whole world is my home!" I tilted my head back and laughed.

"What about your parents? What's their phone number?"

"Did you know that the moon stood still tonight?" I pointed at the sky. "It stopped right in its tracks, and so did everything else! Because God is fire, and God stopped time! Just like She did once before, for Joshua and the Amorites!"

"Really? It stood still?" Dave pointed at the sky too, but with his other gloved hand he held up two fingers. That was obviously a signal to his partner, who'd maneuvered to a position a couple of yards behind me. "You mean that big full moon hanging over those buildings?"

I nodded wildly, my hair flapping. "Yes, it was so beautiful! The light beams froze! They bent and twisted, dozens of loops, all curled around the moon, all perfectly symmetrical! It was God's halo! The frozen fire!"

Dave stepped closer to me but kept pointing at the moon, trying to fix my attention on it. "Whoa, that's amazing." He made the "okay" sign with his other hand. "Dozens of loops, you say?"

Before I could respond, the shorter paramedic came up behind me and grabbed my arms. Dave rushed forward, smooth and choreographed, ready to help his partner restrain me. But I didn't resist, didn't fight them for even a second. Instead, I went limp in their embrace and started crying.

"Oh, God, God! It's the end of all things!"

Then I let out a wail and started mumbling incoherently.

The paramedics flanked me, each gripping one of my arms, and escorted me toward the station's exit. I stumbled between them, head lolling, feet dragging. As we passed the crowd of onlookers, a few people clapped and cheered the paramedics, but most of them had already lost interest. They turned toward the tracks and peered into the distance, looking for the next downtown train.

SirBob1429 kept his camera on me until the paramedics led me out of the station. He laughed again and said, "Only in New York, folks." Then the video ended, and the words "Up Next: Crazy Subway Dancers" appeared on the blank screen.

I clicked on the "go back" button at the top of the screen and returned to the list of "crazy subway" videos. According to YouTube, in just the past five minutes another two hundred people had viewed the Crazy God Girl video. Seventeen of them had also posted comments on it. I had no intention of scrolling through all that junk, but I couldn't help but notice the most recent comment. It was a Bible verse posted by someone named PastorMartin123:

"Then spake Joshua to the Lord in the day when the Lord delivered up the Amorites before the children of Israel, and he said in the sight of Israel, Sun, stand thou still upon Gibeon; and thou, Moon, in the valley of Ajalon. And the sun

stood still, and the moon stayed, until the people had avenged themselves upon their enemies."—The Bible, Book of Joshua, Chapter 10, Verse 12.

I stared at the Bible verse. I'd never read it before. We didn't even have a Bible in our apartment. There was no way I could've come across the verse accidentally.

So how did Joshua and the Amorites get inside my head?

How could I hallucinate anyone saying those words if I'd never heard or seen them before?

* * *

Meanwhile, my parents hovered. There was no need for Mom to stay home—Dad was a freelancer, so he worked in the apartment—but she called in sick anyway and spent the whole afternoon pestering me. Every thirty minutes she knocked on my bedroom door and asked if I needed anything. She clearly didn't trust Dad to take care of me, so she ordered him to stay in their bedroom while she took charge of my recovery.

She fixed me a tuna fish sandwich for lunch, which I ate in bed. Then she tried to start a conversation, a cheery, meaningless talk intended to help me forget about last night. She mentioned her cousin Teresa, the ex-hippie who'd married the Internet millionaire in New Mexico and now wanted to make a big donation to Harvest Team. Mom said Teresa was coming to New York tomorrow to visit one of the soup kitchens that Harvest Team supported. Mom was very excited to see her cousin after so many years, and she invited me to join them on the inspection tour. "It means you'll miss another day of school, but that's not a big deal. Teresa's gonna bring her stepdaughter too. The girl is seventeen, same as you, and this'll be her first visit to New York. So it'll be fun, don't you think?"

I nodded. I knew Mom was just trying to keep an eye on me. She was probably afraid I'd have another "episode" as soon as she let me out of her sight. But I didn't mind. I was in no rush to go back to school.

She left me alone for the next hour, but I could hear her pacing in the hallway and arguing with Dad in their bedroom. I felt pretty certain that they'd seen the YouTube video by now, since it was so easy to find, and it must've boosted their anxiety levels to stratospheric heights. In all likelihood, Mom was already contacting Manhattan's best psychiatrists and making appointments for me at some fancy West Side office. Which was a lot better than Bellevue, but it still scared me. I didn't want to find out how serious my problems were.

Finally, at four o'clock Mom knocked on my bedroom door again. I assumed she was going to start the dreaded discussion of psychological treatment, but instead she seemed bewildered. Her forehead creased and her pretty eyebrows curled. "Joan, there's someone at the door who says he knows you? A boy from your high school named Andrei Mishkin?"

I sat up in bed. I was still wearing my nightshirt. "Andrei? He's *here*?"

"So you know him? He says he has something for you, copies of the notes he took in his classes today."

"Wait a second, how does he even know where I live? I never told him that."

"He says Elena gave him the address. I called Elena's mother this morning and told her you were sick, and I guess Elena must've told this Andrei boy. Is he Russian? He has a thick accent."

I stared hard at Mom. "How much did you tell Elena's mother?"

Mom took a step backward and shook her head. "No, no, don't worry. I just told her you had a stomachache, that's all. And I asked her if maybe Elena could come over here for a visit sometime. Because I thought it might cheer you up."

I clenched my hands. I was trying not to get too angry with Mom, who knew nothing about why Elena and I had stopped being friends. She probably assumed we'd had a silly argument, and now she was taking steps to patch up our friendship, all for the sake of my mental health. So yes, Mom's intentions were good, but in this situation she'd messed things up pretty bad.

"I hate when you do this, Mom. Why do you always have to interfere? Right now the last thing I want is visitors."

She took another step backward, her eyes widening. "Oh God, I'm sorry! I didn't mean to…I'll tell the boy you can't see him. Let me—"

"No, I'll tell him myself. Just give me a minute to get dressed."

Mom nodded. "Okay, okay." Still moving backward, she retreated from my bedroom.

I threw on a pair of jeans and a gray T-shirt with the slogan THIS IS WHAT A FEMINIST LOOKS LIKE. Then I hurried to our apartment's foyer, where Andrei waited in the doorway.

He wasn't dressed in a suit today, I noticed with some satisfaction. He wore khaki pants, a blue polo shirt, and brown loafers—still too formal for school, but a big improvement over yesterday. In his right hand he held an old-fashioned canvas backpack, pale green and frayed, the kind of bag that a soldier in the Soviet army might've carried during the Battle of Stalingrad. He grinned his big dopey grin when he saw me, but his face was flushed and sweaty. He looked like he'd run all the way downtown from Franklin High.

"Joan! I'm so sorry for intruding!" He gave me a quick bow. He also bowed to Mom, who hung back in the living room, watching us from a distance. "I didn't mean to wake you up!"

"No, I wasn't sleeping." I brushed back my hair, patting the mess into place. "I didn't feel so good today, so I stayed home. It's a stomachache, actually."

"Yes, your friend Elena told me." He opened his canvas bag and starting rummaging through it. "When I didn't see you in school, I questioned the other students, and they told me to talk to Elena. She said you were sick and she was planning to visit you, but she couldn't do it today or tomorrow, so she suggested that I go to your apartment instead."

My face grew hot. This was so embarrassing. Elena had lied to him—she was never going to visit me, no matter how much her mother nagged her. So she was probably relieved when the new Russian kid asked her about me. It gave her the chance to foist the duties of friendship on him.

Andrei dug deeper into his backpack. "Ah, here it is!" He removed a sheaf of papers and offered it to me. "I went to the FedEx store and made copies of the notes from my classes. Because you and I are enrolled in some of the same classes, yes? And I thought maybe you'd find the notes useful?"

I stepped toward him. Poor guy. He seemed to have a crush on me. Elena had probably realized this too, so she must've been amused when she sent him my way. I felt a surge of anger, which made my face even hotter, but I forced myself to smile. None of this was Andrei's fault. He was just trying to be nice.

I took the notes. "Thank you very much. That's really considerate of you."

"Wait, I have something else!" He rummaged in his backpack again and pulled out a plastic baggie full of teabags. "This is a special Russian tea, excellent for upset stomachs. I want you to have it, please."

He handed me the baggie. I stared through the plastic at the jumbled teabags inside, which were dark orange and unusually large. "Wow. I don't know what to say. Did you…did you go out and buy these for me?"

Andrei lowered his head, looking sheepish. "No, I already had them in my backpack. I have a sensitive stomach, so my mother makes sure I carry them with me all the time." He closed his bag, then looked up at me. "Well, I'll go now and let you get some rest. I want you to get better as quickly as possible so we can begin our work on string theory!"

Grinning, he held out his right hand. I grasped it and we shook hands very formally. His palm was warm and sweaty.

"Thanks again, Andrei. I'll see you soon."

I held the door open for him as he left the apartment. It was a shame, I thought, that I wasn't attracted to him. He was sweet and good-natured, the kind of guy who would make a great boyfriend. I didn't want to encourage him in that direction—things would get really awkward if he asked me out on a date—but I didn't want to hurt his feelings either. I wanted to be totally honest with him about why I wasn't interested. Because he was a stranger, maybe I could tell him the truth.

As soon as I shut the door, Mom came toward me. She cocked her head and smiled. "Well, well. You never mentioned Andrei Mishkin before."

I frowned. "There's nothing going on."

"All right, I won't pry. He's cute, though. Do you mind if I make that observation?"

"You already made it, so why are you asking for my permission?"

I stepped past her and headed for the living room couch. I picked up the TV remote, even though I had no interest in watching television right now. I just wanted to end the conversation.

But Mom followed me into the living room, under the pretext of adjusting the window shades. She raised them up all the way, then changed her mind and lowered them a bit. Then she leaned closer to the glass and pointed outside. "Look at that. I wonder if that's his father."

"What?" I turned away from the TV.

"Your friend Andrei is in front of the building, talking with someone. A tall man in a gray coat."

I jumped off the couch and rushed over to the window. It was hard to get a good look at them—I was peering straight down at the sidewalk, four stories below—but I could see that Andrei was talking with the same man I'd spotted yesterday on the City College campus.

I shivered as I stared at them. The man had short black hair and a sturdy, solid torso. His face was pinkish and meaty, and he pointed a big, thick finger at Andrei, jabbing the air between them. Andrei said something in response, but I couldn't hear him above the street noise. Then the man grabbed Andrei's shoulders with both hands and started shaking him. The boy's head flopped back and forth like a doll's.

Mom let out a gasp. I opened the window, intending to scream for the police, but before I could start yelling, Andrei twisted out of the man's grasp and ran down the street. The man didn't chase him; instead, he shouted something at Andrei. Although it wasn't in English, I could guess the meaning from his tone of voice. It was definitely a threat.

Then the man walked away in the opposite direction. He moved quickly, his long gray coat flapping around his legs, until he passed out of sight.

Chapter Eleven

At noon the next day, Mom and I took a taxi to the soup kitchen on 60th Street. We could've walked there—it was less than a mile from our apartment building—but Mom was still treating me like an invalid. She seemed to think that any disturbance, even a little exercise, might trigger my next mental breakdown.

The taxi dropped us off at the Church of Saint Paul the Apostle, and we walked down the steps to the church's basement. It was a huge, noisy room full of tables and folding chairs, and at each table eight people were eating lunch. The food was served at a row of banquet tables on the left side of the room, where volunteers in white aprons and paper caps dished out chili and rice from chafing trays. Everything was clean and orderly: the people stood in a neat line to get their food, exchanged pleasant hello's with the volunteers, and waited patiently for seats to open up at the lunch tables. The crowd was a mix of homeless people and down-and-out senior citizens. A couple of old women recognized Mom as we came inside, and they smiled at her, but she didn't notice. She was looking for Cousin Teresa.

I knew the place well, because I used to come here all the time. Harvest Team delivered free food to all the soup kitchens in Manhattan, but St. Paul's was the closest one to our home, so Mom and Dad made it a part of our family life. When Samantha and I were little, we used to volunteer at the banquet tables, ladling pasta and salad on hundreds of plates, smiling at everyone in line. Sammy was my guide and role model then; I always watched her carefully and imitated her. I served the same portions of food as she did, and said the same nice things to the homeless people. When we got older, we helped the cooks in the kitchen, seasoning pots of stew and chopping whatever vegetables had been donated that week. We never went to Sunday services at St. Paul's—or any other church—but we were regulars at the soup kitchen for almost a decade. And though I hadn't visited the place in the past three years, it felt very familiar. It *smelled* familiar: garlicky, cheesy, oniony, citrusy.

After half a minute of futile searching, Mom waved at Father Louis, the assistant pastor who supervised the volunteers. The priest strode toward us, bustling as fast as he could, his potbelly bouncing inside his black shirt, his fleshy neck squeezed by his clerical collar. He raised his arms and hugged Mom, clearly delighted to see her. "Mary! It's been too long!"

She nodded rapidly as they hugged. "You're so right! I miss this place!" She pointed at me, making the priest turn his head. "You remember Joan, don't you?"

"Of course!" Father Louis smiled, but luckily he didn't try to hug me. He was probably scared off by my T-shirt, which said SMASH THE PATRIARCHY. He turned back to Mom. "Listen, your cousin and her stepdaughter arrived a little while ago, but they went to the restroom. They should be back here any minute."

"Oh good. I was wondering where they were. You look wonderful, Louis."

He shook his head. "Please. I've gained ten pounds since July. But you look well, you really do." He stepped closer to Mom and lowered his voice. "I want you to know, Mary, we've all been thinking of you and praying for you, everyone at the church. How are you holding up?"

Mom didn't answer right away. Although the priest hadn't mentioned Samantha's name, he'd evoked her memory, and that was still enough to shut Mom down. Especially here, in this room, which Sammy had loved so much.

"I'm doing okay," she finally said. "Thank you for asking."

But her voice cracked, betraying her. Father Louis winced and fell silent.

Samantha had done much more at St. Paul's than work at the soup kitchen. At the age of fourteen she volunteered to visit the elderly church members in the local nursing homes. At fifteen she started tutoring younger kids in the church's after-school classes, and at sixteen she became an instructor in the English-as-a-Second-Language class for adults. During the summer between her junior and senior years in high school, she went to South America as part of a church mission to build a schoolhouse in a remote Peruvian village. Father Louis had gone on that trip too, and afterwards he said he'd been amazed by Sammy's dedication to the project. That was the subject of the eulogy he'd given at her memorial service last July.

She wasn't a pious God Squad girl or anything like that. She just really enjoyed helping people, much more than I ever did, to be honest. I used to ask her all the time, "Why are you doing this charity stuff? What do you get out of it?" and her answers were always the same: "I love the smile on the old woman's face when I come to visit" or "The little kids laughed so much when I read the story to them," and so on. I used to tell her, "I still don't get it," and then she'd urge me to come with her to the nursing home or the tutoring sessions so I could see for myself. "You could tutor the kids too, you know. Instead of working on your math puzzles all day, you could teach them arithmetic."

But I never did. Samantha was simply higher than me on the moral/ethical/humanitarian scales. She was more giving, more compassionate, more open with friends and strangers. I still loved her, but we were growing apart. I thought she was going overboard with all her good deeds, and I started to wish she were a little less perfect. By the time she went off to college, I'd stopped trying to understand her. We talked on the phone once a week when she was at Princeton, but we weren't best friends anymore. I never told her about my problems at school or anything else that was making me unhappy.

In my daydreams, though, I imagined a closer relationship for us, a return to the tight bond we'd had when we were kids. In the future, I thought, we might live in the same neighborhood or work at the same university. Maybe when we reached our late twenties or early thirties, we could get together on the weekends to have dinner and just enjoy each other's company. And maybe then I'd understand her. Maybe we'd get the chance to become sisters again.

That's what I lost when she died. That chance.

The noise in the church basement faded as Mom and Father Louis stared at each other. The people at the lunch tables continued to dig into their chili and rice, but their conversations ebbed, dropping from a dull roar to a scattered murmur. The awkwardness seemed to fill the whole room. My stomach twisted.

Then a tall woman in a cowboy hat came toward us, threading her way between the tables and folding chairs. She was middle-aged, probably in her fifties, with silver bracelets on her long, slender arms and jet-black hair streaming from beneath her ten-gallon hat. She wore a bright yellow dress embroidered in the Southwest style, with multicolored flowers and vines stitched along the collar. Her face was tanned and a bit leathery from the desert sun, but she'd artfully adorned it with pink lipstick and turquoise earrings.

In an instant, Mom buried her sadness and ran toward her cousin. "Teresa! My God, you look fantastic!"

Teresa struck a pose, swinging her hips to the left. Then she laughed and embraced my mother. "Mary, Mary, quite contrary! Oh, it's good to see you again!"

As they hugged each other, the men and women at the nearby tables shifted in their seats and craned their necks to watch. Mom's cousin from New Mexico really stood out here. She was the reason why so many of the conversations in the room had stopped.

Father Louis tactfully stepped away from us and started talking to one of the aproned volunteers, a bald, muscular man who asked a question about the scheduling of the cleanup shifts. Meanwhile, Teresa pulled back from Mom's embrace and lifted a manicured hand in the air. She twirled her index finger in a circle, a gesture that took in the whole room. "I love this place! There's so much energy here, so much of the life force!" She pointed at the trays of chili on the banquet tables. "And all of this food is donated? It all comes from Harvest Team?"

Mom nodded. "We collect the leftover food from hundreds of restaurants and bakeries and supermarkets, tons and tons of it every day. And our trucks deliver the stuff to soup kitchens and food pantries across the city."

"That's glorious!" Teresa's earrings swayed. "It's like the cycle of life, isn't it? Everything is reused and nothing is wasted. You've done something remarkable here, Mary. You've reestablished the patterns of nature in New York City!"

Teresa was definitely living up to her New Age hippie reputation. She was an unusual kind of hippie, though, a flower child who could afford fancy jewelry and custom-made dresses. According to Mom, Teresa's husband had made millions in the software business. She was probably the richest hippie in all of New Mexico.

Mom put on a smile for her, the same smile she used on all the millionaires who were thinking about giving money to Harvest Team. "Would you like to take a look at the church's kitchen? So you can see how the volunteers feed so many people?"

"I would love that! But wait a second." A distracted look appeared on Teresa's face, and she looked over her shoulder. "Where did Charlotte wander off to? She was right behind me just a minute ago."

"Charlotte? Is that your stepdaughter's name?"

"There she is, over at the food line." Teresa cupped her hands around her mouth and called to her stepdaughter. "Charlotte, what are you doing there? That food isn't for you! Come over here and say hello."

Curious, I looked in that direction. A moment later, I stood face-to-face with the most beautiful girl I'd ever seen in my life.

Charlotte was almost exactly my height, but in every other way she was my opposite. Her hair was long and blond and wavy instead of short and black and ratty. Her skin was as pale as the inside of a seashell, so pale that I couldn't believe she came from New Mexico. And she had no pimples on her face, *none* whatsoever. Her eyes were olive green instead of muddy brown like mine, and her eyebrows swooped across her forehead like wings. Last but not least, she wore a fluttery white dress that bared her shoulders and stretched halfway down her thighs. It was impossible to imagine a bigger difference between her outfit and mine (blue jeans and the SMASH THE PATRIARCHY T-shirt). She looked like a Hollywood starlet, and I looked like her bodyguard.

In her left hand she held a chocolate chip cookie, which she must've grabbed from the tray of desserts at the end of the soup kitchen's food line. While I stared at her, she switched the cookie to her right hand, all five fingers touching its rim. "Hey, you must be Joan, right?"

I just stood there. Seriously, I was afraid to move. I felt like a hiker who sees some beautiful creature in the forest, maybe a deer standing only a few yards away, and she locks eyes with the animal but doesn't move a muscle, because she knows that the slightest movement will break the spell.

Mom gave me a funny look. My silence was puzzling her, and maybe also the expression on my face, which must've been completely abnormal. I felt a bolt of panic—I definitely did *not* want Mom studying me right now. I needed to snap out of it and say something, *anything*.

I nodded at Charlotte, too fast and too enthusiastically. "Yeah, that's right, I'm Joan! Nice to meet you!"

She smiled. She was so pretty, I could hardly stand it.

Her stepmother frowned. She pointed at the chocolate chip cookie in Charlotte's hand. "You weren't supposed to take that." Her earrings jiggled in disapproval. "The food here is for the homeless people."

Mom shook her head. "No, no, it's okay. There's plenty for everyone."

Still smiling, Charlotte broke the cookie in two. Then she stretched her hand toward me, offering one of the halves. "Want to share?"

I was very aware of Mom and Teresa watching us. I knew I shouldn't get so paranoid about it; we were just sharing a cookie, that's all. It was totally normal, as wholesome as a scene in a Disney movie. And yet I felt their eyes on me as I took the crumbling morsel from Charlotte's hand. "Uh, thanks."

To my infinite relief, Mom finally turned away. She scanned the room and tracked down Father Louis, who'd just wrapped up his conversation with the bald, muscular volunteer. Taking the priest's arm, Mom steered him toward us. "Louis, I want to give my cousin a tour of the kitchen and then go over some details about Harvest Team's budget. While we're doing that, could you take the girls to the sanctuary upstairs? I bet Charlotte would love to see some of the gorgeous artwork up there."

Louis hesitated. He clearly disliked the idea of babysitting us. But his soup kitchen relied on Mom and Harvest Team, and Mom wanted some private time with Teresa. So after a second he nodded. "Of course. It would be my pleasure." He put on a fake smile and turned to Charlotte and me. "Come on, girls. I'm going to show you the loveliest church on the Upper West Side."

He led us out of the church basement while Mom and Teresa headed for the kitchen. Charlotte walked a couple of feet to my left and took a bite of her cookie. I faced forward but strained my eyes all the way to the left so I could watch her. I thought again of a beautiful forest creature, stretching its neck toward a tree branch and nibbling the leaves.

Without any warning, she stepped closer to me and brushed her shoulder against mine. "Aren't you going to eat your half?" She pointed at my right hand, which had closed over the cookie. "It tastes a little stale, but it's not so bad."

"Oh yeah, I forgot!" Flustered, I opened my hand. The cookie had broken into smaller pieces and the chocolate chips were melting. Embarrassed, I popped the whole thing into my mouth to get rid of the mess as quickly as possible.

Charlotte laughed, but not in a cruel way. She seemed genuinely amused. "Whoa, impressive. You're a fast eater."

My mouth was full, which gave me a few seconds to decide how to respond, but my mind stayed blank as I chewed the cookie. The soup kitchen's baked goods were always stale—they were donated by bakeries after sitting on the shelves for several days—but they didn't lose any of their sweetness. The chocolate melted on my tongue, and I couldn't say a word.

Father Louis looked over his shoulder at us and frowned. "Come on, keep up! There's a lot to see."

* * *

The church basement and the main sanctuary had separate entrances on the street, the former directly below the latter. Father Louis led us outside and we climbed the steps to the sanctuary's front doors. Before we went inside, though, the priest pointed at the bas-relief mural carved into the stone above the doorway.

"This artwork shows the conversion of Paul, the patron saint of our church. He's the figure on the right side, the one falling off his horse and stretching his arms toward the sky." Father Louis swept his hand to the left. "And the figure on the left side is God, who spoke to Paul with such force that the heavenly voice knocked him to the ground. The Lord converted Paul to Christianity and commanded him to spread the faith to the whole world." He paused, looking at us as if we were his Sunday-school students. "Charlotte, are you Catholic? Maybe you know this story already?"

She shook her head. "No, I belong to the Church of Everlasting Light."

Father Louis raised an eyebrow. "I never heard of that denomination. Is it new?"

"Well, my dad started it five years ago. After my mom died, he sold his software company and got more involved in spiritual things." Charlotte ran a hand through her hair and twined a blond lock of it around her middle finger. "That's how Dad met Teresa, actually. She heard about Dad's church and started coming to the meetings at White Sands."

The priest smiled, but it wasn't sincere. There was something mean about it. "White Sands? Is that the town where you're from?"

"No, it's a desert in New Mexico. A really beautiful, spiritual place. Our church used to hold its meetings there. Usually at night, because it was too hot during the day." She shrugged, lifting her bare shoulders. "But my dad got sick last year and then he couldn't run the church anymore. So now it's an Internet-only thing. We have a really cool website and thousands of followers."

Father Louis let out a snort. "Sure, that's the trend now, isn't it?" His tone was mocking, sarcastic. "Everything else is on the Internet these days, so why not religion too?"

Charlotte scrunched her eyebrows, confused, probably wondering why this priest was smirking at her. But I didn't have to wonder. I knew all about Father Louis and his two-thousand-year-old church. Its arrogance was the main theme of my "History of the Middle Ages" class.

I stepped past him and opened one of the church's doors. "We better hurry, Charlotte. We need to check out Christianity's greatest hits before the Internet puts it out of business."

She laughed. It was a beautiful sound, soft and pattering, like rain in the forest. She came toward me and strolled through the doorway, her white dress fluttering around her legs. Father Louis followed her, unamused. His smirk was gone, and that made me very happy.

Once we stepped inside the church, though, I stopped feeling so cheerful. It was a huge, intimidating space, almost as big as a football field, the ceiling a hundred feet high. There were Gothic arches and statues everywhere, and enough pews for a thousand people, although right now there wasn't a parishioner in sight. The place was dark too, an urban cavern. Because the church was surrounded by tall buildings, very little sunlight came through the stained-glass windows. There were racks of votive candles everywhere, but weirdly enough, their light seemed to intensify the darkness. At the far end of the sanctuary, a marble altar supported a golden crucifix, which gleamed faintly in the candlelight.

I felt cold. I hugged myself, rubbing the backs of my arms. After Samantha died, Father Louis had offered to hold the memorial service here, but my parents asked him to come to the funeral home instead. I think the church scared them just as much as it scared me.

The priest headed straight for the baptismal font, which was spectacularly grand. A pedestal that looked like a drinking fountain held a small basin of holy water, which flowed in an arcing stream to a much larger basin below, an irregular octagon that was as big as a hot tub. Ignoring us, Father Louis dipped his right hand into the water and crossed himself. A few yards away, Charlotte tilted her head back and peered at the high ceiling, her mouth open. Meanwhile, I shivered and rubbed my arms harder, thinking of Saint Paul and the heavenly voice that had knocked him down. The old Bible story stirred a new uneasiness in my stomach. It was too close to my hallucinations.

Then I heard footsteps in the darkness. I spun around and saw a pitch-black figure emerge from the shadows. It was a looming silhouette, otherworldly, terrifying.

I stopped rubbing my arms. I stood there, motionless, my heart thudding. I was hallucinating again. I was sure of it.

No! Not here! Not again!

Father Louis turned around and stepped toward the figure. "William, I'm glad to see you. Do you have a minute?"

It wasn't a hallucination. It was just another priest, a tall man in a black shirt, a little younger and thinner than Father Louis but not so different otherwise. He gave Louis a deferential nod. "Sure, what's on your mind?"

I stared at the man for a few more seconds, just to confirm he was real. Then I stumbled toward the baptismal font, my legs shaking. I leaned against the pedestal for support and took a couple of deep breaths.

Father Louis glanced at me. He was probably worried that I would desecrate the holy water somehow. "Joan? Can I leave you and your friend alone here for a while? I need to talk with Father William about something."

I nodded weakly. "Yeah, sure. No problem."

"I'll be back soon. In the meantime, you can look at all the artworks in the side chapels." Father Louis grasped the other priest's arm and they headed for the church's offices.

Charlotte stopped gazing at the ceiling and watched the two men leave. Then she looked at me. She scrunched her eyebrows again. "Hey? Are you all right?" She stepped toward the baptismal font. "Whoa, you're really pale. Do you feel sick?"

I was nauseous and shivering, but most of all I was embarrassed. This was absolutely the worst time for me to fall apart. "Yeah, sorry. I've been kind of… under the weather lately."

"Oh, you poor thing." Charlotte approached the pedestal and wrapped her arm around my shoulders. "You're cold too." We stood side by side, facing the basin of holy water, like a couple of elementary-school girls taking turns at a drinking fountain. "Here, let me warm you up."

She started rubbing my arm, her hand sliding up and down between my elbow and shoulder, her fingers slipping inside the armhole of my T-shirt. This warmed me up pretty effectively, although it wasn't the friction that did it. When she touched my skin I felt a burst of adrenalin, a small hot explosion in my stomach. It wasn't a pleasant feeling, exactly—it was so intense it almost made me squirm. The heat spread to my chest and arms and legs, and in a few seconds I stopped shivering. I wasn't nauseous anymore, but I was jangled and jumpy and not thinking straight. I couldn't believe this was happening.

This isn't just a friendly touch. There's something more to it.

After a while Charlotte stopped rubbing me and gave my shoulder a final pat. "Feeling better? I bet you could use a drink of water." She pointed at the basin. "Is it okay to drink holy water, you think? Or would it be a sin?"

She smiled, and I couldn't really tell if she was serious. The curves of her upper and lower lips were slightly mismatched, giving her a mischievous, teasing look.

I smiled back at her, trying to mimic her expression. "I don't think it's a sin, but I wouldn't drink it. All the parishioners stick their hands in that water, and some of them aren't so hygienic."

"So it's for external use only?" Charlotte winked at me and dipped both of her hands into the basin. She held them just below the surface, palms up, fingers wiggling. "You know, in the Church of Everlasting Light we also used water in our services."

"You mean the church your dad started?"

"Yeah, Dad wrote a bunch of new prayers for our church, and a lot of them were about water and baptism. Hold on, let me see if I can remember one." Charlotte cupped some of the holy water in her palms and held it above the basin. Her wet hands glistened in the candlelight. "Okay, here goes: 'May the water of life fall on every corner of the desert. May it bring forth the ecstatic flowers of harmony and cosmic consciousness.'" She laughed and spread her fingers, letting the water spill back into the basin.

I laughed too. Charlotte's giddiness was contagious. "That's pretty deep. So did you participate in the services too? As an altar girl?"

"Oh yeah, I loved it. We were way out in the desert in the middle of the night, so we could do anything we wanted. We danced and sang like crazy. I don't want to brag or anything, but I have a pretty good singing voice. I used to write my own prayers and belt them out across the sand dunes."

"Sounds like fun. I wish I'd been there."

"No joke, I was the Taylor Swift of White Sands. I was the Lady freakin' Gaga of New Mexico." She laughed again, still leaning over the basin of holy water. "I recruited new members for the church too. That's why I got kicked out of my high school."

"What? Seriously?"

"Yeah, the principal was this stuck-up Presbyterian. He said I was corrupting my fellow students with fanatical ideas."

"Wow, that's..." I didn't know what to say. I was truly impressed. Charlotte was a radical, a religious revolutionary. "So what happened then? Did you find another school?"

"Nah, Dad home-schooled me. At least for a while. After he met Teresa, he didn't have as much time for it. He fell hard for her." Charlotte stood up straight and dried her hands on her dress. "The truth is, though, Teresa kind of ruined our church. At first she was really into it, went to all the meetings, tried to write her own prayers. But then she started complaining that the services were too long. She didn't like the dancing or singing either."

"Really? I thought she was a hippie. Or one of those New Age people. Don't they love dancing around? You know, with flowers in their hair?"

Charlotte frowned, pressing her mismatched lips together. "For Teresa, the New Age thing is just a fashion choice. I mean, she talks the talk, the Earth-is-our-mother mumbo-jumbo, but at heart she's a pretty boring person. The only thing she really enjoys is spending Dad's money."

She shook her head. All her giddiness had vanished. Clearly upset, she lowered her gaze and stared at the basin.

I was tempted to put my arm around her and try to comfort her the same way she'd comforted me. But I didn't. I was too nervous. All I could do was lean a little closer to her. "You don't get along with your stepmom?"

She thought it over for a moment, then shrugged. "I don't know. Things might've gotten even worse if Dad hadn't married her." She kept staring at the holy water. "I mean, Dad's really sick now, too sick to do anything. So if I didn't have a stepmom, they might've put me in a foster home. Or sent me to live with my grandmother, which would've been a total disaster."

"What's wrong with your dad? Is it cancer?"

Instead of answering, she turned around and stepped away from the baptismal font. "Let's go see the rest of the church. Maybe there's something interesting over that way." She headed for the row of chapels on the left side of the sanctuary.

I hurried to catch up with her. Charlotte obviously didn't want to talk about her dad. Which made sense, I guess—we were still strangers, after all. We'd met just fifteen minutes ago, so it was a little early to start confiding in each other. And yet there was something familiar about Charlotte, something I'd imagined and anticipated for a long, long time. So I was impatient. I wanted to get close to her right away, *right now*.

Side by side, we walked past the chapels. Each was tucked in a dimly lit alcove and set apart from the rest of the sanctuary by a knee-high marble railing. One chapel had a statue of the Virgin Mary; the next had a painting of Jesus being baptized; the next had a couple of confession booths. Charlotte glanced at each altar, but she didn't stop for a closer look at any of the statues or paintings. She didn't say anything either, not a single word. She seemed kind of annoyed, like a shopper at a supermarket who can't find what she's looking for and gets more and more irritated as she walks down all the aisles.

We finally stopped at the last chapel on that side of the church. On the wall was a painting of a bearded man in a white robe, kneeling beside a Roman soldier who held a big sword. The man in the robe had a halo over his head, but he looked too old to be Jesus. His beard and eyebrows were white. There were two sentences written in gold letters on the painting's frame—"I have fought a good fight" and "I have kept the faith"—but unfortunately they didn't explain what was happening in the picture.

I nudged Charlotte with my elbow. The silence was making me antsy, so I'd thought of a way to break it. I was going to ask her if she knew who the old guy in the painting was.

But when I looked at her, I noticed she was crying. She wiped her cheek with the back of her hand.

"I'm sorry, Joan. I can't pretend anymore. Teresa told me not to say anything, but I know what's going on." Her voice was a miserable whisper, so soft I could barely hear it. "Teresa was talking to your mom on the phone yesterday, and I overheard what they said. About what happened to you two nights ago."

I froze. This was bad. It was horrendous. "How much do you know?"

Charlotte stared at the floor. Her long hair veiled her face. "Your mom said you had a breakdown. The paramedics found you in a subway station and took you to a mental hospital. And someone put a video of it on YouTube."

My heart sank. She knew the whole story, pretty much. "Did you watch the video?"

"Yeah. I'm sorry about that too. I couldn't believe what I overheard, you know? So I had to see it for myself." She raised her head and looked me in the eye. "But I want to tell you something. You shouldn't be ashamed of it. Not one bit."

I turned away from her. I stared at the painting of the old bearded man.

Charlotte stepped sideways, moving between me and the painting. She was determined to get my attention. "No, I mean it. Most people are too blind to see this kind of thing, but I watched that video a dozen times and I saw what really happened to you. It wasn't a breakdown. It was a revelation."

I looked past her, still studying the painting. I focused on the colorful figures in the background, a crowd of ancient-times spectators in tunics and togas. They clasped their hands and knelt in prayer behind the old man and the Roman soldier.

"Joan, it happened to me too. Three years ago, during one of our church services in the desert. From the outside, it must've looked like I went nuts. I was jumping around and waving my arms and shouting all kinds of gibberish. But it felt different on the inside, way different. It was the most amazing thing I've ever experienced."

Charlotte stepped closer, so close I couldn't ignore her. She'd stopped crying, but her blond eyelashes were still wet. She looked at me with such focus and intensity that her eyes seemed to get brighter, their green irises shining even in the dim chapel. She wasn't playing games. She was serious.

I stared back at her, our faces almost touching. I needed to know more. "What did you see when that happened? When you were jumping around and shouting?"

"It was a glow. A beautiful, warm happiness." Charlotte smiled and raised her head, gazing at the church's ceiling again. "I felt like I was bathed in the divine spirit."

"But this, uh, this divine spirit...did you actually *see it*? Can you say what it looked like?"

"It wasn't really a visual thing. It was more of a *feeling*, you know?"

I nodded and gave Charlotte a grateful look, pretending to be satisfied with that answer. But in reality I was disappointed. Her revelation was nothing like mine. It was joyful and pretty and not at all psychotic. She never saw God as a homeless woman in a subway car, and the Lord never commanded her to solve the equations of string theory. So she couldn't really help me.

To my surprise, though, Charlotte seemed to sense my disappointment. She stopped smiling and grabbed my right hand. She squeezed it as she stared at me, furrowing her brow, looking worried. "But here's the problem. From the outside, no one can see the beauty or the happiness. Your mom can't see it, that's for sure."

"My mom? What—"

"She doesn't know what to do about you, so she asked Teresa for help. That's what I overheard when they were on the phone yesterday. Since Teresa was coming to New York anyway, they agreed to talk about it in person once she got here. I guess that's why your mom sent us up here with the priest, so she could have a private conversation with Teresa downstairs."

I felt a twinge of fear. My forehead ached and my throat tightened. "But... but how could Teresa help?"

"She has connections at New Life Oasis. It's a place in New Mexico, near Santa Fe. It looks like a health spa, because it's so fancy and expensive, but it's really a mental health center." Charlotte grimaced. "Teresa knows everyone at New Life because that's where my dad is. She checked him into the center a year ago, and he's been there ever since. That's what I meant when I said he was sick."

I pulled my hand out of Charlotte's and stepped backward. My throat was so tight, I could barely speak. "So my mom…she wants to…she's trying to put me…"

"I mean, don't get me wrong, it's one of the best treatment centers in the whole country. All the Hollywood celebrities go to New Life to work out their problems. And the celebrities' kids too." She shook her head. "It hasn't helped my dad, though. They gave him drugs to calm him down, but they didn't cure him. And sometimes they give him too many drugs, and he acts like a zombie. He hardly says a word when I go there to visit."

I took another step backward. This was beyond embarrassing—it was crushing. How could Mom do this to me behind my back? How could she make arrangements at this treatment center without even consulting me? I raised my hands and clamped them on either side of my forehead, trying to smother the blazing pain inside my skull. I'd been getting headaches for the past two days, ever since I woke up in Bellevue Hospital, but this one was the worst yet.

Charlotte reached for me again, but I backed away from her. I was humiliated. I couldn't believe I was the last one to know about this. "So…when will it happen? When will I have to go to this…this…"

"No, you should fight it! They can't force you to go there." Charlotte narrowed her eyes and gave me a fierce look. "If you want, I can help you. Because of my dad, I know all about the procedures for checking into New Life. Your parents will have to send you to a psychiatrist in New York first, and then the psychiatrist will have to send a report to the center, so that means you'll have a couple of days to try to stop—"

Footsteps echoed, coming from the front of the church. When I turned that way I saw Father Louis striding toward us. He'd finished talking with the other priest, and now he was ready to resume his babysitting duties. He spread his arms wide, as if he were delighted to come back to us.

"Hello again, girls. I see you've found our finest treasure." He stopped in front of the painting of the old man and the Roman soldier. "What do you think of it?"

Charlotte and I exchanged glances. It was a silent agreement. We would return to our conversation as soon as we could. We would put our heads together and make a plan.

Father Louis waited a few seconds for an answer, then pointed at the painting. "Its title is 'The Martyrdom of Saint Paul.' It shows the end of the story that I started telling you outside, the final chapter in the life of our patron saint. The Romans beheaded Paul for spreading the Christian faith."

I didn't want to look at the painting again, but my eyes drifted toward the Roman soldier's big sword.

Louis raised his hand and made a chopping motion. "The execution took place outside the walls of Rome. According to the legend, after the executioner chopped off Paul's head, it bounced three times in the dirt, and three fountains sprang up in the places where it hit the ground."

I forced myself to look at the old man in the painting. Now I realized what the sentences on the picture frame meant. Saint Paul the Apostle had fought for God. He'd kept his faith till the end.

But I didn't have any faith. When I tried to see where my own story was going, dread churned inside me.

This wasn't going to end well.

Chapter Twelve

An hour later, I was back in my bedroom. Which was starting to feel like a jail cell.

Dad was my guard now. Mom had dropped me off at the apartment, then hurried over to the Four Seasons Hotel on 57th Street, where Teresa and Charlotte were staying. But before Mom took off, she gave Dad an unpleasant job that she clearly hadn't wanted to do herself. He came into my room and told me about my appointment the next day—at one o'clock on Thursday afternoon—with Dr. Peter Cauchon, clinical psychiatrist.

"He's the best, Joanie. He went to Harvard Medical School. And he's won all kinds of awards." Dad sat down next to me on the edge of my bed. He opened his laptop so he could show me Dr. Cauchon's website. "It says here that 'he specializes in helping teenagers navigate the joys and disappointments of young adulthood.' Sounds good, right?"

I stared at the site's home page. It displayed photos of smiling teens who looked more like fashion models than psychiatric patients. The accompanying text was equally cheerful and carefully avoided any upsetting words such as "schizophrenia" and "bipolar disorder."

Frowning, I turned away from the laptop and looked at Dad. "So this psychiatrist is going to explain what happened to me? Why I freaked out on the subway?"

"Yes, but he'll use language that's a little more scientific. Instead of 'explaining why you freaked out,' he'll make 'a clinical diagnosis.' That's the more sophisticated term, FYI."

Dad's attempt at humor fell flat. I was in no mood for jokes. "What if something is seriously wrong with me? What happens then?"

"Let's not get ahead of ourselves, okay?" His voice was quiet, calm, reassuring. "When you see the doctor tomorrow, he'll ask you some questions and try to figure out what's going on in your life. And maybe he'll schedule a follow-up appointment so he can monitor you over the next few weeks."

"Okay, but you haven't answered my question. What if it turns out to be something really bad? Like schizophrenia?"

The word jolted him, no doubt about it. Dad blinked a few times and swallowed hard. After a couple of seconds, though, he composed himself and shook his head. "Look, you're jumping to conclusions. In all likelihood, this was a one-time thing. You were tired and stressed and that's the whole story. But even if I'm wrong, Joanie, even if the problem is bigger than that, we'll deal with it, all right? We'll figure out the options and find the best treatment for you."

He reached for my shoulder and squeezed it. Dad was trying to show his love and support, but it all felt a bit hollow, because I knew he wasn't being completely honest. He and Mom were already preparing for the worst-case scenario. That's why Mom had talked to Teresa about the New Life Oasis treatment center.

But I wasn't going to confront him about it. If I did, he and Mom would realize that Charlotte had tipped me off, and they'd probably make sure I never talked to her again. And I didn't want that to happen.

So I just nodded. "Okay. I'll stop worrying about it."

That was what he wanted to hear. He smiled and stood up. "In the meantime, why don't you work on some math puzzles? That'll help take your mind off your troubles, right?"

I nodded again. Satisfied, Dad left the room.

Afterwards, I lay in bed and stared at the ceiling for a while. Under ordinary circumstances, I would've taken up Dad's suggestion and reached for my math notebook, but my headache had gotten worse since I returned to the apartment. The pain in my temples sharpened whenever I tried to concentrate, making it hard to work on any kind of math problem and completely impossible to think about string theory and Calabi-Yau manifolds. I'd never had such a bad headache before, so I assumed it was an aftereffect of my breakdown. My brain was rebelling against its hallucinations. It was defying the illusory homeless-woman God who'd ordered me to solve the mathematical mystery of the universe.

Then, as I rubbed my temples, my iPhone vibrated. I pulled it out of my jeans pocket and saw a text from a telephone number I didn't recognize.

I'm worried about you, my new friend and fellow mathematician. You've been gone from school two days now. Are you still ill, Joan? Did you try the tea I gave you yesterday?

It was Andrei. He must've gotten my phone number from Elena. Out of the five hundred kids at my high school, he was the only one who'd noticed my absence. Clearly, his crush on me wasn't going away. If anything, it was getting heavier. We needed to have a serious talk about my sexual orientation, preferably soon, but I couldn't broach this topic with a text. The best I could do right now was send him a polite reply.

Yeah, still sick. But its headache now, not stomach. Thx for thinking of me.

His next text arrived within seconds.

That Russian tea is good for headaches too!! It's very soothing. Should I come to your apartment again to give you copies of today's class notes?

I felt as if a very friendly dog was jumping up and down to get my attention. It was sort of cute and sort of annoying.

No, dont worry. I'm gonna take a nap now and forget about school. Thx again!

I thought this might put an end to the conversation, but Andrei's next text arrived just as fast.

That sounds like a good idea! But you can call me tonight at this telephone number, especially if you want to talk about string theory. I'm going to stay up late to study the subject, so you can call me anytime!

He was obviously trying to pique my interest and get me to respond, but I put the phone back in my pocket. Although Andrei seemed like a nice enough kid, he was also awkward and intense, and I was still a little freaked by what I'd witnessed the day before, his scuffle on the street with the tall man in the gray coat. Mom had assumed the man was Andrei's father, but I didn't think so. He'd looked more like a gangster, a thug who had no qualms about thrashing a skinny teenager. That was something else Andrei and I needed to talk about, preferably soon. But like the sexual-orientation talk, we couldn't do it by text.

A minute later, though, my iPhone buzzed again. Reluctantly, I pulled it out of my pocket to see what Andrei had written, but this text was from a different phone number.

Hey, Charlotte here. I got your # from your mom. She went with Tere$a to have drinks in the snobby bar downstairs (my $tep-monster loves to get $mashed) but I stayed in the hotel room. How r u doing?

I jumped out of bed, holding the phone like a stick of dynamite that might explode at any second. I'd hoped that Charlotte would get in touch, but now that she'd texted me, I didn't know how to respond. Should I be casual or serious? Funny or thoughtful? Nonchalant or emotional? I had no idea, but I knew I had to text *something* before too many seconds passed, so I jabbed at the keys on the iPhone's screen.

Thx for getting in touch! I have an appointment with a psych doctor tomorrow. The good news: I get to take another day off from school. The bad: they might ship me off to the New Life center. Dad says my problem isn't that serious, but he's probly lying. He's too scared to tell me the truth.

Charlotte wasn't as fast at texting as Andrei was. I could tell she was writing a reply, because a row of gray dots streamed across my iPhone's screen, but I had to wait almost a whole minute before her text popped up.

Be careful!! Specially if the doc asks questions about RELIGION!!!! You'll get in trouble if you say you believe in anything besides the "normal" religions. I'm serious, I have experience with this!! The doctors said my dad was crazy because of the new church he started.

I felt a spasm in my stomach, a surge of anxiety. When Mom and Dad had set up the appointment with the psychiatrist, they'd probably told him about my breakdown and what I was ranting about when the paramedics found me. So there was a good chance that the topic of religion would come up during my appointment.

What should I say if the doc asks me about it? I can't deny what happened at the subway station.

This time Charlotte was quicker.

Tell him it was just a joke. Just say you were angry at your parents or your school (or whatever) but no one paid any attention to you. So you decided to make a big scene and start yelling about God. Because you knew it would get everyone worried.

I shook my head as I read the text. That strategy wouldn't work. I could never pull off such a complicated acting job. But I didn't have any other ideas, and I also didn't want to sound ungrateful, so I chose to be diplomatic.

OK I'll try it. Thx for your advice, Charlotte. Even tho we just met a few hours ago, I feel like I can trust you.

I was probably going too far, saying too much, but I hit the SEND button before I could stop myself. Then I was in agony for the next thirty seconds as I waited for Charlotte to reply.

You're special, Joan. I hope you dont mind but I told some of my friends about you. These are people I met on the Everlasting Light website. I've been running the site and the chat group ever since my Dad got sick.

I loved the first sentence of her text. It made me feel warm and light, as if I'd just stepped into a shaft of sunshine. But the second sentence brought me crashing down. How many people had Charlotte talked to? And what exactly did she say about me? I was so nervous as I wrote back to her that I kept pressing the wrong keys. It took me forever to correct all the spelling mistakes.

OK cool, but you didnt mention my name, did you? This whole thing has been so embarrassing.

I suffered another half-minute of agony.

I didnt say your last name. I just said "my friend Joan." Believe it or not, some of the Everlights (thats what I call my chat group friends) already knew about you. They saw your video on Youtube.

I felt another spasm. This one was so painful, it made my eyes water. The iPhone's screen blurred, but I managed to tap out a question.

You mean the video of me at the subway station?

It was a pointless question—there were no other videos of me circulating on the Internet—but I had to ask it. I had to confirm the worst.

Yeah its going viral. Its getting lots of views because it feels so true. People get chills when they watch it. They see right away that your experience was the real thing. So everyone wants to share it.

Trembling, I pressed the iPhone's home-screen button, then clicked on the Web browser and went to YouTube. The phone kept me waiting for five horrible seconds, and then it displayed the page for "Crazy God Girl in NYC Subway Station."

When I'd checked the page yesterday, the video had five thousand views. Now it had almost two million.

I felt hot all over, boiling hot, incinerator hot. I tried to calm myself by thinking about math—prime numbers, complex functions, partial differential equations—but the only thing I could focus on was a probability formula. Making conservative assumptions about the rate of YouTube viewership in New York City, I tried to estimate the percentage of kids at my high school who'd seen the video.

But I gave up on the calculation before figuring out the answer. I had a feeling that I was never going back to Franklin High anyway.

I pressed my phone's home-screen button again. Charlotte had sent me another text.

Hey Joan? U still there?

My hands were shaking.

Yeah I'm good. Just tired. Gonna lie down for a while.

* * *

I slept for the next five hours. In the 48 hours since my breakdown, I'd slept almost twice as much as I normally did. Like my headaches, I think this was my body's natural reaction to the craziness. My waking life had turned insane, and my only escape was to lose consciousness.

It was 9 p.m. when I got out of bed and staggered down the hall to the toilet. Mom and Dad intercepted me as soon as I got out of the bathroom, and they coaxed me into the dining room for a late supper of chicken and rice. While I picked at the chicken, Mom regaled me with amusing stories about her reunion with Teresa, telling me how spacey her cousin was, and how many margaritas she'd downed at the hotel bar, and how much money she planned to donate to Harvest Team. Dad dutifully laughed at the funny parts and asked me three times if I wanted another drumstick. My parents were putting on a show for me, acting as if this was just an ordinary Wednesday night, as if I were going to school tomorrow and not to an appointment with the most intimidating psychiatrist in New York City.

It was exhausting.

By the end of the meal, my head was splitting. I took four ibuprofens, but they did nothing to ease the pain. So in desperation I decided to try Andrei's tea. I filled a mug with boiling water and immersed one of the crumpled teabags, which turned the water a strange shade of orange. The taste was even stranger—like a salty apple—but I forced myself to drink it. Then I returned to my bed and rubbed my temples and lay there until midnight.

I couldn't sleep now. The headache was like a fire inside my skull, burning the crisp tissue behind my eyebrows. It ignited my thoughts, scorching and tormenting me. How many people were viewing me on YouTube at that very moment? Had Elena seen the video yet? Or Professor Taylor or Principal Barnes? What about Father Louis, was he a secret YouTube junkie? Was he watching me rant about God right now and wondering how I knew so much about the Amorites?

The pain and agitation finally drove me out of my room. Across the hall, my parents' bedroom was dark, but they were light sleepers, so I was careful not to make a sound. I tiptoed through the living room and dining room and kitchen, navigating the convoluted layout of our apartment, which my parents had renovated before I was born, back in the days when they had more money and hope. Then I came to the apartment's third bedroom, a cramped nine-foot-by-eight-foot space on the other side of the kitchen. It was really too small to be a proper bedroom; a real-estate agent would probably recommend using it as a home office instead. But it had been big enough for Samantha.

Originally this room had been our playroom. When Sammy and I were little, we both slept in a bunk bed in the larger bedroom and used the smaller room for our Legos and jigsaw puzzles. But seven years ago, when Samantha was twelve and I was ten, Mom and Dad decided we should have separate bedrooms, and to everyone's surprise, Sammy offered to take the smaller room. She said she liked it because it was cozy, and though this preference seemed absurd to me—the larger bedroom was better in so many ways—I took her statement at face value and reveled in my good fortune.

It was only later that I saw her choice as part of a lifelong pattern. Samantha believed in self-sacrifice. Unlike most people, she enjoyed giving more than taking.

I stepped into the room, but my head was pounding, so I didn't turn on the light. The moon shone through the window, illuminating the narrow bed and small desk and swivel chair. There wasn't enough space for a bureau, so Sammy had put up shelves in her closet and kept all her clothes there, neatly folded and arranged. Her desk was neat too: the only things on it were a lamp, a clock radio, a couple of pencils, and a tube of ChapStick. These items rested in the same places where Sammy had left them last May, during the week between the end of her freshman year at Princeton and the start of her summer teaching job in Peru. Mom and Dad never went into this room anymore, so nothing inside it had changed.

I stood by the window and looked outside. The room didn't have a good view; it faced the backs of two apartment buildings on the same block. But a fifty-foot gap separated the two buildings, and the moon hung in the slender strip of night sky between them. It was waning now, two days past full. The darkness crept in from the right, stealing across the bright disk.

The sight of it was enough to trigger a panic attack. My heart started racing and my head throbbed with pain. I remembered what had happened the last time I stared at the moon, how the homeless woman had commanded it to stop in its orbit, and how its light had frozen into a warped halo, fantastically twisted and intricate. I closed my eyes, but I could still see the brilliant beams looping across the sky, threading into and out of the darkness.

It was like fire, yes it was. The flames leaping and dancing. The very same fire that killed my sister.

I let out a whimper. The pain was too much. I was going to throw up. I was going to black out again.

But I shut my eyes tighter and bit my tongue to stay conscious. Because I wasn't just scared—I was *angry* too, shaking with fury. I was angry at my broken brain for hurting so much, and for hallucinating the homeless-woman God, and for leading me into my dead sister's bedroom. And I was angry at

Samantha for getting herself killed. It was so *stupid* the way she'd acted, all that giving and sacrifice. She should've stood up for herself when she was twelve years old. She should've asserted her rights as the older sister and demanded the larger bedroom.

And when she was in Peru last summer, teaching English and math in the village of Nauta, she should've stayed away from the schoolhouse after it caught fire. Even though there were children trapped in the building, she should've stayed outside. She should've let it burn.

I wanted to scream. I was in so much pain.

But instead I opened my eyes, leaned across Samantha's desk, and grabbed the tube of ChapStick. In a frenzy, I twisted off its top and attacked the window, scrawling long streaks of lip balm across the glass. I used the ChapStick like a crayon and copied the pattern I saw so clearly in my head, drawing a twisted, sinuous halo on the window, a ring made from dozens of smaller loops, the whole thing curving around the blazing moon. It took me a couple of minutes to sketch the entire halo and scribble all its twists and turns. When it was done, the greasy streaks gleamed on the glass, shining in the moonlight like a ring of frozen fire.

Then a miracle happened. While I stared at the warped halo I'd drawn, my headache went away. I laughed as the pain eased, I was so relieved and grateful.

At the same time, I realized three important things:

One, I needed to clean the window right away. If my parents woke up and saw the crazy picture I'd just scribbled on the glass, they wouldn't wait for the results of my appointment with the famous psychiatrist. They'd call the New Life Oasis center first thing in the morning and put me on the earliest flight to New Mexico.

Two, I couldn't clean the window, because a piece of Samantha was on it now. She'd used that tube of ChapStick before she left for Peru. She'd left traces of her DNA in the lip balm, and now they were part of the halo.

Three, I recognized the pattern I'd drawn. It was a rough sketch of a Calabi-Yau manifold. It showed a new way to fold up the extra dimensions of string theory.

I stared at it in wonder. It was a completely different kind of geometry, wild and unexpected. The loops were the holes in the structure, specifying the shape of the manifold. The space was warped in a way I'd never seen before, and yet the configuration had a logic to it, a satisfying rightness. It was beautiful. It was perfect.

I tiptoed back to my bedroom, grabbed my math notebook, and brought it to Samantha's room. Then I picked up one of the pencils from her desk and began to calculate. This new manifold had so much geometrical symmetry that the calculations were surprisingly easy to do. The complex equations of string theory boiled down to a simpler set of formulas that showed how the microscopic strings would vibrate inside those folded dimensions. And if this manifold was truly the correct one, the formulas would spit out the magic numbers of our universe: the strengths of all the forces and the properties of all the particles.

Before I even finished the first calculation, though, I sensed that I had the correct solution. Out of the almost infinite number of possibilities—10 to the power of 272,000—I'd found the true shape of the universe.

But was it really me who found it? How could I have guessed the right solution from the $10^{272,000}$ choices? That would've been ridiculously improbable. The chance of making the correct guess was microscopic, less likely than the probability of a monkey typing the complete text of *Hamlet*. It couldn't have happened that way.

No. Someone had given me the answer. Someone had drawn the shape for me in the night sky above 125th Street, during what I thought had been a hallucination.

Which meant I hadn't really hallucinated the homeless woman on the subway. Or the boy lying on the trail in Van Cortlandt Park. I realized now that I hadn't imagined them, because they'd told me a secret I couldn't have discovered on my own.

I saw the truth, the universe's most astonishing fact.

God is real.

Chapter Thirteen

An hour and a half later, at 2 a.m., I slipped out of my apartment building and flagged down a taxi heading south on Broadway. I told the driver to go to Second Avenue and Tenth Street, in Manhattan's East Village, and I settled into the backseat as the cab sped downtown. I knew the ride would be expensive, at least twenty-five dollars, but I was in a hurry. Like most mathematicians, I never believed anything unless there was a rigorous proof for it, and I felt the same way about what I'd experienced tonight. I needed to confirm my discovery of God.

The taxi cruised down the empty streets at thirty miles per hour, a speed that would've been unthinkable in daytime traffic. In fifteen minutes we reached the destination, and I gave the driver nearly all the cash in my wallet. I got out of the cab, clutching my math notebook to my chest, and stepped toward Samovar, which was the name of the 24-hour diner on the corner.

Even at 2:15 a.m. on a Thursday morning, people were still on the streets of the East Village. Down the block, a crowd poured out of a dance club called The Yellow Brick Road, which was decorated with big rainbow flags, their wooden poles jutting from metal stands on the sidewalk. I loved seeing those flags and watching the couples come out of the club, especially the pairs of women. They laughed and held hands as they strolled down the street. Most of the clubbers were young, only four or five years older than me, yet they seemed so much freer and cooler and happier. Several of them headed for Samovar, which appeared to be their late-night after-party hangout. Through the diner's windows I saw half a dozen tables occupied by the clubbers: boys with man-buns and leather pants, girls with shaved heads and short skirts and tight tops.

But just outside Samovar was their polar opposite, a boy from the other end of the coolness spectrum. Andrei Mishkin stood by the diner's entrance in his khakis and polo shirt, holding a stack of textbooks under his arm. He came toward me and grinned his big dopey grin.

"Joan! You're here!" He didn't bow this time, but he tipped his head forward a bit. In public settings I guess he modified his usual greeting to make it a little more discreet. "I'm so happy you called me. This is wonderful news."

"Yeah, I'm sorry about waking you up, but—"

"Don't worry, I'm not tired now. How could I be tired after hearing about this breakthrough of yours? You didn't have any trouble getting here, did you?"

"No, it was easy. What about you? You live close by, right?"

"Just a few blocks away. Luckily, I managed to leave the apartment without waking up my mother. She won't get up till seven, so we'll have plenty of time to talk." He gestured at the textbooks under his arm. "I was also lucky to get these books about string theory from the library yesterday. Maybe they can help us analyze your new findings. And Samovar has excellent food, so we can eat while we work."

I walked with Andrei to the diner's entrance. "Yeah, I like the look of this place. There's a 24-hour diner in my neighborhood too, but it's kind of depressing."

He opened the door for me. "Samovar is so famous, we heard about it as soon as we came to New York. They serve simple, good Russian food. My mother and I have eaten here many times already." We passed the diner's lunch counter and an old-fashioned grill where a fat, bearded cook fried rows of yellow dumplings. "Let's go to one of the tables in the back. It's quieter there."

We walked past the clubbers, who didn't bother to look at us. At the back of the restaurant all the tables were empty, so we chose a big one in the corner and set up our workspace. Andrei dumped his textbooks on the table and pulled a stubby, chewed-on pencil from his pocket. I opened my math notebook and turned the pages until I found my pencil sketch of the new Calabi-Yau manifold. It was an exact copy of the one I'd drawn on Samantha's window.

Before I could show it to Andrei, though, an elderly waitress in a white smock shuffled over to our table. Her gray hair was elaborately braided, with the twisted locks curled around her head like a crown, and the name *Kateryna* was stitched into her smock. She beamed at Andrei, wrinkling her face in delight, obviously recognizing him. "Andrusha! You're here so late! Are you doing homework?" She pointed at the string theory textbooks. "Or studying for a test?"

He smiled back at her. "Yes, Katya, a very important test."

"And is this one of your friends? From your new school?" The waitress examined me for a moment, narrowing her eyes, then turned back to Andrei. "My God, why didn't you start studying earlier? How will you wake up in the morning if you stay out so late?"

"Your food will give us all the energy we need. We'll have two bowls of borscht, yes?" Andrei glanced at me. "You'll love the borscht here, it's delicious. And we should have some pelmeni too. You like potato dumplings, Joan?"

"Sure, I'll try them." In fact, I was pretty hungry. I hadn't eaten much for dinner. "But I have to warn you, I don't have a lot of money on me right—"

"No, no, this is my treat." He turned back to Kateryna. "We'll have the borscht and the pelmeni."

The waitress smiled again, but this time it was a sly, knowing smile. Leaning closer to Andrei, she whispered something in Russian. Andrei blushed and muttered, "*Da, da.*" Kateryna laughed and walked away.

I frowned. I didn't like being left out of the conversation. "What did she say?"

His blush deepened. Andrei's face turned as pink as a pencil eraser. "She said you're very pretty."

I looked down at my shapeless jeans and the T-shirt I'd thrown on twenty minutes ago, the one that said THE FUTURE IS FEMALE in big black letters. Either the waitress was joking, or her eyesight was terrible. But Andrei seemed to believe her. He did his best to act casual—he reached for his math notebook, which looked just like mine, with a spiral binding and a blue cover—but his ears were reddening. Although I'd arranged this meeting to discuss string theory, Andrei clearly saw it as something more. In his eyes, it was our first date. Why else had he offered to pay for everything?

I shook my head. I wanted to be honest with Andrei, but I was too wimpy to tell him the truth. He'd missed the hints I'd tried to give him about my sexuality, including the one written on my shirt. Or maybe he was in denial, just like my parents, unwilling to acknowledge the signs. So my only option now was to come out and say it: "I'm a lesbian." It was the natural thing to do, the logical thing.

God wants me to come out to Andrei, I thought. *God loves me for who I am. And He—no, She—wants me to be proud of myself.*

This thought popped into my brain, completely out of nowhere, and yet it felt so right it made me shiver. It made sense: If God were real, why would She hate gay people? Why would She stop them from loving each other? Only a perverse, made-up God would act that way. The God I'd glimpsed was nothing like that.

But no, I can't come out to him yet. First we need to prove the hypothesis that God exists. Once we answer the Big Question, all the little ones will be easy.

And the crux of that hypothesis was string theory. I believed that God had shown me the solution to a virtually unsolvable problem, a puzzle I could've never cracked on my own. But I wasn't a hundred-percent sure yet that my solution to string theory was correct. There was always the chance that I'd made a stupid mistake in my math or missed something obvious. So I wanted Andrei to perform the same calculations and see if he came up with the same results.

I handed him my math notebook, held open to the pencil sketch of the manifold. "Here's what I mentioned on the phone, the new kind of Calabi-Yau space. I calculated how string theory would work if the extra dimensions were folded up this way. All my calculations are on the next few pages, but I don't want you to look at them, all right? I want you to do the calculations yourself and double-check my work."

He nodded, his eyes focused on the sketch. He looked at it with curiosity at first, and then with growing wonder. For the first time since I met Andrei, he seemed transfixed, at a loss for words. He placed my notebook on the table in front of him, handling it with care. Then he picked up his pencil, opened his own notebook, and started filling its pages with equations.

Neither of us spoke for the next half hour. Andrei chewed his lower lip as he worked, applying the formulas of string theory to the new shape I'd discovered. His stubby pencil jiggled in his hand, writing line after line of numbers and symbols. Every so often he stopped and squinted at what he'd written, looking over his math. His eyes darted back and forth between my sketch of the manifold and his scribbled calculations.

I watched him carefully. I thought of the purity of mathematics, how it clarifies the world, how it shines right through all our ugly thoughts and reveals the beautiful order behind them. And I thought of the ecstatic axiom

I'd shouted at the paramedics during my breakdown at the 125th Street station: *Our great task is to understand the cosmos, and our best tool is mathematics.*

After ten minutes Andrei's eyes widened and he let out a surprised grunt. He leaned over his notebook and attacked the problem with renewed energy, and after another ten minutes a look of amazement spread across his face. I knew what had stunned him, because I'd experienced it myself: the equations of string theory were simplifying before his eyes. When the formulas were applied to the new manifold, the intractable terms and infinities canceled out. The equations became streamlined and much easier to solve.

By this point Kateryna had brought the bowls of borscht to our table, but we didn't touch them. Andrei threw himself into the final calculations, hunched over his notebook, oblivious to everything else. He was breathing hard, almost panting. Sweat beaded on his forehead.

Finally, he wrote the last formula with a flourish. "Okay! I've found a value for alpha!" He turned his notebook around so I could see what he'd written. "Is it correct? Is it the same as the measured value?"

My heart leapt when I saw it.

0.00729735

Alpha was the fine-structure constant, which specified the strength of the electromagnetic force in the present-day universe. Andrei's value for it was the same as the one I'd computed. More important, it was the same as the value that had been measured by scientists in thousands of experiments. Using my new manifold and the equations of string theory, we'd calculated one of the universe's magic numbers. And we could use the same procedure to compute all the other physical constants—gravitational, nuclear, cosmological. The equations generated the observed features of nature, all the properties of atoms and stars and galaxies, tying everything together. They were the blueprints of the cosmos.

My eyes stung. I blinked a few times, trying not to cry, and then I looked at Andrei. "The number's correct. It's exactly the right value. Like hitting a bull's eye."

"So the theory works? It can predict the strengths of all the forces?"

"Yeah, I think so. This is it, what everyone's been looking for. The Theory of Everything."

He nodded, slow and serious. Then he reached for a glass of water and took a long drink.

Weirdly enough, we didn't celebrate. We didn't cheer or high-five each other or jump up and down. Instead, we started eating our borscht. I think we both needed some time to think about it, to contemplate the magnitude of what we'd just done. I stared at my bowl of beet soup, purple and thick, with a dollop of sour cream and a sprig of dill on top. I watched myself dip my spoon into the soup and bring it to my lips.

After a couple of minutes, Andrei put his spoon down. He'd stopped sweating, and his breathing had come back to normal. "It's hard to believe, yes? Are you as shocked as I am?"

"Most definitely. And you know who's going to be even more shocked? Professor Laura Taylor."

He let out a squeaky laugh. "What do you think Professor Taylor will say when we tell her we solved the puzzle of string theory?"

"At first she'll say we're wrong. She'll say we made a mistake somewhere." I smiled. "Once we prove we're right, though, she'll probably try to take credit for it."

"Yes, that would be typical."

"But we shouldn't tell her anything just yet. We have a lot more work to do. We need to calculate *all* the constants to make the proof airtight. And that's going to take a while."

"You're right, we're going to be busy. But before we proceed, there's one thing I'd like to know." Andrei pointed at my sketch of the Calabi-Yau manifold. "How in the world did you figure out the correct shape for the extra dimensions? This manifold is so complex, so intricate. Did something inspire you?"

The short answer was yes, so I nodded. But I wasn't sure how to explain the rest. I closed my eyes and saw the number 137—so very close to the reciprocal of alpha—on the chest of the runner in Van Cortlandt Park. *Should I start the story there? Or should I skip ahead to the subway car and the twisted halo around the moon? How can I tell him the truth without sounding crazy?*

"Joan? Are you okay?"

I'd kept him waiting too long. I opened my eyes. "Listen, I need to ask you a personal question. Do you believe in God?"

Andrei leaned back in his chair. "Really? You want to talk about religion?"

"If it's too personal, you don't have to—"

"No, no, just give me a moment." He raised his hand to his chin and tapped his index finger against his lips. He was clearly thinking about the question, and that made me happy. He was taking me seriously. "I have a confused background when it comes to religion. My parents never sent me to church or religious school, but I think I inherited some of their beliefs anyway."

"What do you mean?"

"Well, my mother came from a small village where everyone belonged to the Russian Orthodox Church, but she gave up her faith after she moved to St. Petersburg. My father, on the other hand, was a devoted Communist, a true believer in Karl Marx. He was full of socialist idealism when he married my mother. But then Communism collapsed in Russia, and that devastated him." Andrei looked down at the table. "All this happened more than ten years before I was born, so I didn't see it myself. But Father once told me that he was very depressed for many years afterward."

"I'm sorry to hear that. He got better, though?"

"Yes, eventually he changed his focus and started working for a computer company. By the time I was born, my parents were tired of belief. But they still liked to tell me stories, so I grew up with some fondness for God and Marx."

This was interesting. There were similarities between Andrei's background and my own. But he hadn't really answered my question. "Okay, I'm not talking about fondness for the idea of God. Do you believe God exists, really and truly?"

Andrei thought about it some more. He scratched his pale cheek. "We're mathematicians, yes? So we need to define our terms. How do you define God?"

"The creator of the universe. The designer of the laws of physics, the prime mover behind the Big Bang." I thought of what the homeless woman had told me in the subway car. "An intelligence that still exists as an unseen influence and may even guide the evolution of the cosmos."

Andrei leaned forward, resting his elbows on the table, and stared at me. His expression was a mix of curiosity and concern. "Excuse me, but I think I see where this is going. Are you saying that *you* believe in God?"

"Well, I just think we should keep an open mind about—"

"And you believe God inspired you to choose that particular Calabi-Yau manifold?"

His voice was heavy with incredulity. I guess I should've expected it—all mathematicians are infuriatingly logical. But Andrei looked at me with so much skepticism at that moment—cocking his head, raising his eyebrows—it just seemed a little excessive. It was as if I'd said my revelation had come from the tooth fairy.

How can I convince him that I'm not crazy? That God might've actually shown me the way?

Because Andrei was a mathematician, I figured the best strategy would be to present a mathematical argument. So I took a deep breath and reached for one of the string theory textbooks. I was going to find the chapter on Calabi-Yau spaces and show him the estimate of $10^{272,000}$ possible manifolds. Once he saw the numbers, he'd have to admit the crucial point: it would be astronomically improbable to guess the right manifold without any help.

But before I could make this argument, someone approached our corner of the diner. He grabbed a chair from one of the nearby tables, dragged it across the floor, and placed it next to Andrei's chair. Then he lowered his massive torso onto the seat, without taking off his long gray coat.

"Hello, Andrei. You seem surprised to see me here."

It was the tall man I'd seen on the City College campus and in front of my apartment building. The thug who'd grabbed Andrei by the shoulders and shook him like a doll.

He ignored me. He didn't even glance in my direction. He leaned forward and stuck his big, meaty face in front of Andrei's. "You shouldn't be surprised, my friend. I warned you about this." He curled his lip on the last syllable, baring his teeth. "I told you what would happen if you broke the rules."

Chapter Fourteen

When the man finally turned to me, he leered. His lips were thick and wet, his teeth yellow and uneven. His eyes were dark slits.

"And hello to *you*, Miss Joan Cooper. Yes, I know your name, and the address of your family's apartment too. Two-one-five West Seventy-Eighth Street, correct?"

His accent was thicker than Andrei's. His voice had the rasp of a heavy smoker, and his coat smelled of cigarettes. Underneath it he wore black pants and a black silk shirt. His shoulders bulged under his coat, and his muscular neck strained his shirt collar, but what scared me the most were his hands. They were large and sharp-knuckled and laced with crooked scars.

He saw my fear. His wet smile widened. "You're not so bad-looking, Little One. You should dress better, though. And comb your hair."

I turned away from him and looked for help. I needed to get the attention of one of Samovar's employees. Craning my neck, I spotted Kateryna at the other end of the diner, but she was taking orders from a table of clubbers.

He kept leering at me. "You know, if you tried a bit harder, you could be attractive. Which makes me wonder why you're spending time with our ugly friend here." He glanced at Andrei. "How did you convince this girl to go out with you? Did you promise her something?"

Andrei shook his head. All the color had drained from his face, and his eyes were a pair of frightened ellipses. He swallowed hard, and his Adam's apple bobbed in his throat. "Please. Just leave her alone."

The man shrugged. The smell of cigarette smoke grew stronger. "Sorry, it's too late for that. I warned you, didn't I? About sharing information?"

"We're not—"

"Shut up. I'm already sick of talking to you. I'm going to talk to your girlfriend now." He turned back to me, his mouth half-open. I could see his fat tongue pressing against his lower lip. "So, Little One, did Andrei ever tell you about me? His old Uncle Vlad from St. Petersburg? To be honest, I'm not actually his uncle, not in the biological sense. But I used to work very closely with his father."

At that moment Kateryna turned our way, and I nearly jumped out of my seat to catch her eye. She came toward us. "What is it? You want to order something else?"

I pointed at the thug who'd called himself Vlad. "This man came out of nowhere and sat down at our table and started harassing us. Could you please ask him to leave?"

The waitress glared at him. Her wrinkled face turned fierce. "Who are you? Why are you bothering these young people?"

Vlad leered at her too and said something in Russian. I had no idea what the words meant, but they sounded harsh and foul. Kateryna opened her mouth but didn't say anything. She took a step backward, then another, gaping in shock at the man, as if he'd just threatened to kill her whole family.

Then she spun around and hurried to the front of the diner. She found another elderly waitress and the bearded cook, and the three of them huddled by the grill, talking in frantic whispers. But they made no move to call the cops. They seemed afraid to even look in our direction.

Vlad chuckled and turned back to me. "They may not know my face, but they recognize my name. You see, I have a reputation among the Russians who've settled here in New York. They know I come to America every few months or so. And when I'm here, they know enough not to interfere with my work."

"So you're a gangster?" I stared right at him, refusing to be intimidated. "You're in the Russian Mafia?"

He let out a dismissive snort. "Don't insult me, Little One. I'm associated with a much more powerful organization."

"What? The Russian government?"

"You don't need to know the details. Let's just call it a committee." He pointed at Andrei. "This boy's father worked for a computer company that did projects for my committee. He wrote software for us, very important and effective computer programs. And because this boy is a mathematical prodigy, his father taught him programming and sometimes showed him the company's software."

Andrei leaned across the table and gave Vlad a desperate look. "Please stop! She knows nothing about this!"

"Oh really?" Vlad raised one of his bushy eyebrows. "So why are you sharing all this technical information with her?" He pointed at the pair of math notebooks lying open on the table. "These equations? I'm no expert on computer science, but I think I can guess what they are. They're part of the last program your father wrote for us, yes?"

"No!" Andrei rose from his chair. "This has nothing to do with my father! These are physics equations, not computer programs! You have it all wrong!"

Vlad frowned. "Sorry, I don't believe you. Just two days ago I saw both of you go to the mathematics department at City College to talk with one of the professors there. And I know that your father's last program involved some very complicated mathematics." He clenched his hands as he stared at Andrei. The scars on his knuckles reddened. "Don't try to tell me it's a coincidence. I know a lie when I see it."

I stood up too. I had only a vague idea what was going on, and I couldn't understand why Vlad was so worried about a computer program. But I knew two things for sure: Andrei was in trouble, and he needed my help. So I bent over the table and tapped a finger on the formulas he'd written in his notebook. "Andrei's right. This is string theory. See the differential forms? The tensor products? Those are terms that appear in physics equations." I moved my hand toward my own notebook and pointed at the sketch of the Calabi-Yau manifold. "And see this figure? It's about as far from a computer program as you can get. I don't know what you're accusing Andrei of, but you're making a mistake if you think—"

Without any warning, he rose to his feet and slapped me.

His thick hand walloped the left side of my face, hitting me so hard that I saw a white flash beneath my eyelids. My head jerked to the right, and I had to grab the corner of the table to stop myself from crumpling. My ears rang and my vision blurred. My legs felt like jelly.

"Enough talk, Little One. Now I'm getting sick of your voice too."

The whole diner saw it. The clubbers stopped eating, the cook stopped frying dumplings, and the waitresses gasped and covered their mouths. Vlad scowled at them all, daring them to challenge him, but no one did. An awkward paralysis spread across the room as they stared at the big Russian, everyone frozen in shock. None of the clubbers stood up or shouted at Vlad or called 911. With just one blow, he'd terrorized the place.

He turned back to Andrei. "Did you forget the conversation we had in St. Petersburg? On the day after your miserable father killed himself?"

Andrei backed up against the wall. He raised his arms in surrender. "Please…please don't—"

"I agreed to let you and your mother leave the country, but there was one condition. Whatever you knew about the software, you swore to keep it secret. You promised not to tell *anyone*."

"I haven't! Those notebooks are all physics! Nothing else!"

"We'll see about that." Vlad picked up both notebooks from the table. "I'm going to show these equations to the experts on my committee. They're smart people, so they'll know if you're lying. And if you are, you'll be in a lot of trouble, my friend."

The big Russian held the notebooks under his arm. I was still wobbly, still holding on to the table for dear life, but I realized what was happening. Vlad was stealing the Theory of Everything. Even though he was a brutish idiot who didn't know the first thing about math or physics, he'd seized our solution—the solution God had shown me!—and planned to hand it over to his equally brutish accomplices. And I knew with all my heart that I couldn't let that happen.

I took a deep breath and narrowed my eyes, focusing on Vlad's obscene face. Then I stood up straight and stepped toward him.

"Give them back." I pointed at the notebooks in his hand. "Give them back *right now!*"

Vlad cocked his head, clearly amused. He pulled his arm back and eyed me as if I were a mosquito. "What, are you a masochist? You want another taste?'

He tried to slap me again, but I was smarter this time. I jumped backward, out of reach, putting at least six feet between us. Then I pointed at him and stomped my sneaker on the floor. "*This man assaulted me!*" I shouted loud enough for everyone in the diner to hear me. "*You all saw it! Someone call the police!*" I looked around the room, glancing at all the stunned faces. "*This isn't a joke! I need help!*"

Neither the waitresses nor the cook moved a muscle, but several of the clubbers rose from their chairs. A girl in a miniskirt reached for her cell phone, and two dudes in New York Knicks basketball jerseys stepped toward our table. Vlad glared at them, but they kept on coming. In the meantime, three more guys approached us from the other side of the diner.

I stomped the floor again. I felt a strange new confidence and power, a surge of unbridled strength in my arms and legs. It was like the feeling I usually got at the end of a cross-country race, the burst of energy for the final sprint, but now it spread through my whole body and prickled my skin. I kept pointing at Vlad. "*That's right, stop him! See those notebooks in his hand? He's trying to steal our work!*"

Vlad scanned the room, apparently weighing his options. He glanced with disdain at the young men coming toward him, then looked over his shoulder at Andrei, who was still cringing against the wall. Then he looked at me and let out a sigh. "Ah, that was a mistake, Little One. You're going to regret it."

Scowling, he turned around and headed for the exit, with the notebooks still in his right hand. The dudes in Knicks jerseys tried to intercept him, but he extended his left arm and shoved them out of his path. While the clubbers yelled at him and the waitresses shook their heads, he rammed the front door open and barreled out of the diner.

Without a second thought, I followed him outside. I wasn't going to let him get away with it. He had no right to steal what God had given me.

* * *

Vlad turned south after leaving Samovar. He strolled briskly down Second Avenue with the notebooks under his arm, tilting his head back and gazing at the 3 a.m. sky. As I walked thirty feet behind him, I heard him whistle a familiar, cheery tune. It surprised me a little that he knew it, but I guess certain songs are popular all over the world, even in Russia. It was "Don't Worry, Be Happy."

The street wasn't quite as lively as it had been an hour ago, but plenty of folks still crowded the sidewalk. The Yellow Brick Road was still going strong, and at least twenty people were hanging out in front of the dance club, chatting and smoking cigarettes next to the long row of rainbow flags. Standing beside the entrance were the club's bouncers, a pair of tall, burly bruisers with impressive pectoral muscles bulging inside their black T-shirts. They stood by the door with their rippling arms folded across their massive chests. One man was white and the other was Asian. Each was just as big and intimidating as Vlad.

I ran over to them. "Hey, guys, I need your help!" I pointed at the big Russian, who'd reached the corner of Second Avenue and Ninth Street. "That man stole something from me!"

The white bouncer didn't react at all. The Asian guy unfolded his arms and gave me a blank stare. "Uh, what was that?"

"He also hit me!" I pointed at my left cheek. It was still stinging from the slap, so I assumed it looked reddish or bruised. "I want him arrested for assault!"

The bouncer furrowed his brow. "Look, we're not cops. You should call 911."

By this point, Vlad had crossed Ninth Street and was more than a hundred feet away. I felt a stab of frustration. "A girl at the diner already called the police, but he's getting away!" I clutched the bouncer's thick forearm and tried to pull him down the block. "That's why I need you! You have to grab the guy and hold him until the cops get here!"

"No, listen, I'm sorry." The bouncer shook his head and wrenched his arm out of my grasp. "I'd like to help you and all, but we gotta stay by the door."

I stomped my sneaker on the sidewalk, but this had no effect on the bouncer. He folded his arms across his chest and went back to his job of doing nothing at all. The dance club paid him to guard the door, but all he really had to do was look tough. Unless someone tried to sneak into the club or break its rules, he would just stand there like a statue all night.

Fuming, I looked down Second Avenue. Vlad was two hundred feet away now, almost out of sight. He was disappearing down the street, taking away my solution, my Theory of Everything.

Then it hit me. I needed to break the rules.

I went down the sidewalk in front of the club until I reached the last rainbow flag at the end of the row. Its wooden pole was five feet high, and it rose from a flag stand in which the bottom end of the pole rested. Putting one foot on the base, I grabbed the flagpole and yanked it out of its stand. Then I waved the rainbow flag over the sidewalk to get the bouncers' attention.

"Look at me, suckers! I'm stealing your flag!"

"HEY!" The Asian bouncer sprang into action. He stepped forward and pointed at me. "PUT THAT DOWN!"

"No way! Come and get me!"

Then I turned around and started running.

I dashed to the corner and crossed Ninth Street, then sped down Second Avenue. Peering down the street, I caught a glimpse of Vlad's bulky silhouette about a block ahead. Better yet, I heard furious footsteps behind me. The Asian bouncer was chasing me down the sidewalk.

"STOP! STOP RIGHT THERE!"

His voice boomed down the street, echoing against the shuttered storefronts. And Vlad heard it too. He stopped at the next corner and looked over his shoulder at us. I raised the rainbow flag high above my head as I sprinted toward him.

"You heard the man!" I shouted. "Stop right there!"

If it had been just me chasing him, Vlad would've laughed and slapped me again, probably a lot harder this time. But the Asian guy was spectacularly ripped. The sight of him charging down the street would've been enough to make *anyone* panic. So after a moment of hesitation and disbelief, Vlad did the sensible thing. He faced forward and bolted down Second Avenue.

I smiled as I ran after him. The Russian didn't seem so threatening now. He turned his head to the left and right as he careened down the street, looking for an escape route, his long coat flapping and fluttering behind him. Clearly, he wasn't built for speed, and he wasn't wearing the right shoes for running either. He couldn't keep up this pace for very long. But I was a Varsity cross-country runner, one of the best in the whole city. I could go on like this for hours.

I tightened my grip on the flagpole, holding it straight up with both hands. I felt that surge of confidence and power again. It flowed through me like the wind.

Vlad was a hundred feet ahead, his shoes clomping the sidewalk, and the bouncer was a hundred feet behind me. I thought the Russian might turn left or right at the next street corner, but luckily the next street was Saint Mark's Place—the heart of the East Village—and it was crawling with late-night partyers. A guy in a New York University sweatshirt saw us first; he shouted, "Hey, look at them go!" and all the people around him craned their necks to watch us.

Without breaking stride, I lowered the rainbow flag until the pole was horizontal, its tip pointed at Vlad. "*Stop him! He's a criminal! He's a thief and a spy!*"

But no one came forward to help. Instead, they raised their fists and cheered and turned their heads as we ran past. They didn't think it was serious. New York is full of crazy spectacles, especially late at night, and the crowd of partyers must've thought they were watching some weird parade or performance. Several of them held up their phones to take pictures.

I grimaced. I couldn't count on anyone coming to my rescue. There were no police officers or patrol cars in sight, and if the Asian bouncer caught up to us, he was more likely to side with Vlad than with me. I had no allies in this fight. I was alone.

Then I felt a new burst of strength in my legs. It coursed through me like an electric shock, like a revelation.

I'm not alone. God is on my side.

I ran harder, faster. I hurtled down Second Avenue, leaving Saint Mark's Place far behind. I steadily closed the gap between Vlad and me, and by the time we reached Sixth Street he was only ten yards ahead. He was close enough that I could see the sweat on the back of his neck, gleaming under the streetlights. I ran so fast that my sneakers barely touched the sidewalk, and I held the flag in front of me like a knight with his lance.

I was God's soldier now. I was invincible.

"Stop, Vlad! The Lord Almighty commands it! This is the will of God!"

My voice rang to the heavens. It was so loud and fierce and triumphant that no one on the street could ignore it. All the partyers on Second Avenue tilted their heads to listen, and all the sleepers in the nearby apartment buildings sat up in their beds. And Vlad turned his head too, he couldn't help it. He looked over his shoulder in confusion and surprise.

Then he tripped. The tip of his shoe struck one of the crevices in the pavement, and he lost his balance. He toppled forward, his arms pinwheeling, his solid torso falling like a two-hundred-pound bomb. His chest hit the sidewalk and slid across its rough surface, stopping only when his head slammed into a fire hydrant.

In two seconds I caught up to him. I didn't know at first if I'd have to attack him or start doing CPR, but after a moment Vlad let out a groan and sat up next to the hydrant. He raised his hands to the back of his head, which glistened with blood, but luckily he didn't seem to be dying. At the same time, though, he looked too groggy to fight, which was a big relief.

Best of all, he'd dropped the math notebooks. I stooped to the sidewalk to pick them up, then backed away from the groaning Russian.

The bouncer arrived a second later and pulled the rainbow flag out of my hand. But before he could start yelling at me, a New York Police Department car came barreling down Second Avenue, flashing its lights and blaring its siren.

The car veered toward us and pulled up to the curb. I wanted to run again, but the bouncer had a firm grip on my arm.

I was in trouble now.

Chapter Fifteen

Only one cop came out of the patrol car, but he was impressive. I could tell from his uniform that he was an NYPD boss, a police captain. He wore a white shirt with a badge on the chest and gold bars on the collar. He was in the department's top ranks in terms of body weight as well. His potbelly ballooned the front of his shirt.

He emerged slowly from the police car, hefting his torso out of the driver's seat and loping to the curb. He adjusted his officer's cap and examined the scene on the sidewalk: the big Russian sitting next to the fire hydrant and rubbing his bloody scalp, the bouncer holding the rainbow flag in one hand and gripping my arm with the other. Although the cop wasn't a paragon of physical fitness, he looked pretty intimidating just the same. His lips were pressed into a tight, hard frown, and his eyes were small and grim. The nameplate below his badge said CARPENTER.

He pointed at Vlad first. "Don't get up, sir. Just stay where you are, okay?" He had a cop's commanding tone and a strong Brooklyn accent. "I'm gonna call an ambulance for you."

Vlad stared at the police captain for a couple of seconds. Then he put on an expression I hadn't seen on him before, a friendly, helpful, accommodating smile. "Thank you for your concern, officer, but I'm not badly hurt." He'd changed his voice too, making it softer and more pleasant. "I don't need an ambulance."

"Really?" Captain Carpenter squinted at him. "You got a deep cut there. Looks pretty bad to me."

"No, it's not serious. It hardly hurts at all." Vlad pulled a handkerchief out of his coat pocket and pressed it to his wound, smiling all the while. "I'm more embarrassed than anything else."

"So what happened here?" Carpenter kept his eyes on Vlad, most likely because he appeared to be the victim. Of the three of us, he was also the best-dressed. "Was there an altercation of some kind? I was driving by and saw you fall down."

Vlad shook his head. "No, no, it was just an accident. I heard someone running behind me, and I got startled. So I started running too, and then I tripped and fell."

I was confused. Vlad clearly hated my guts, so why wasn't he blaming me for his injury and demanding my arrest? Was he trying to downplay the incident in the hope that the cop would go away? Because the Russian wanted nothing to do with the local police?

Captain Carpenter seemed puzzled too. Frowning, he turned to me and the bouncer. "What about you two? Why were you running down the street at this time of night? And what were you doing with that flag?"

The bouncer tightened his grip on my arm. "This kid stole the flag from the Yellow Brick Road. You know, the dance club. I work the door there." With his free hand, he pulled an identification card out of his pocket. "So I ran after her. The girl is crazy, if you ask me. An underage nutjob."

Carpenter glanced at the bouncer's card, then focused on me. "What's your name, kid? And how old are you anyway?"

My stomach started churning. When the paramedics took me to Bellevue two nights ago, they probably reported the incident to the NYPD. Which meant I couldn't give my name to the police captain now, because he'd check the records and see the report. Then he'd take me back to the psychiatric ward, and this time it wouldn't be so easy to get out.

I gave him a desperate look. "Uh, sir? I'm not the one you should be worried about." I pointed at Vlad, who was slowly rising to his feet. "That man is a Russian spy! He threatened a friend of mine and hit me in the face!"

Unfortunately, this just made things worse. Carpenter stared at me. "Let's take it one step at a time, okay? Right now I just need your name."

"I'm telling you the truth! It happened at Samovar, the diner up the street. A girl there called the police, so you should check with 911 or something. Or go back to the diner and talk to—"

"Listen, I need some cooperation here." The cop leaned closer. "If you don't give me your name, I'll have to take you to the station house. Is that what you want?"

I said nothing. This wasn't good.

In the silence that followed, Vlad stepped toward Carpenter. "FYI, I'm an art dealer, not a spy. And I've never been inside that diner." He reached into the pocket of his coat, pulled out a card, and offered it to the police captain. "Would it be possible for me to go home now, officer? I need to disinfect all these scrapes and cuts. You can reach me at this number if you have any more questions."

Carpenter took the card and nodded. "All right, you can go. But make sure you take care of that gash on your head."

"What? You're letting him go?" I glared at the cop. "Don't you see what's happening? That card he gave you is a fake! Now he's gonna go back to the Russian embassy or wherever, and you'll never see him again!"

The captain ignored me. While Vlad stepped around the corner and made his escape, Carpenter turned to the Asian guy. "Well, you got your flag back. You want to press charges against this girl?"

"Nah, I gotta get back to work." He pushed me toward the cop, who grabbed hold of my arm just as the bouncer let go of it. "Just keep her away from the dance club, all right?" He gave me a final look of disgust, then headed back to the Yellow Brick Road.

Meanwhile, the cop led me to his patrol car. With one hand he opened the car's rear door and with the other he nudged me into the backseat. He wasn't rough about it, but he wasn't easygoing either. He closed the door behind me and locked it.

I leaned back against the seat and shook my head. This was a disaster, the worst possible screw-up. No matter what I said or did, I was going back to Bellevue. The only bright spot was that I'd recovered the math notebooks, which I still clutched in my right hand. But what good were they going to do me in the mental hospital?

Carpenter opened the driver-side door and sat down behind the steering wheel. Through the steel grate that separated the car's backseat from the front, I could see his fleshy neck and the graying hair under his officer's cap. He turned the key in the ignition and put the car into drive, but before lifting his foot off the brake he looked at me in the rearview mirror. I saw his small dark eyes in the rectangle of glass.

"You can relax now. I'm taking you home."

I didn't understand. "Home?"

"Yeah, back to the Upper West Side. Your address is two-one-five West Seventy-Eighth Street, right?"

"Wait a second, how do you know that?"

"I took a new shape tonight, Joan. A man this time, instead of a woman or a boy."

He stepped on the gas, and the car pulled away from the curb, but the dark eyes in the mirror focused only on me. They glowed like bits of onyx, like miniature black suns. And as I stared at him I realized he was only pretending to be a cop. If I had a list of all the captains in the New York Police Department, I wouldn't find anyone named Carpenter on it.

I was looking at another disguise for the Almighty.

He nodded, confirming my guess. "Sorry about all the drama back there, but it was necessary. We need to talk again."

Chapter Sixteen

The police car cruised down Second Avenue. The driver's eyes gazed at me from the rearview mirror. "Want me to turn on the air conditioning? You look like you're hot."

If I were in the car with a real cop and he'd asked me the same question, I would've probably said, "Yeah, sure, turn it on." I was still sweating from my sprint down the street, and the back of the car was pretty warm. But I didn't say a word. I was stunned and disoriented and very, very confused. The man driving the police car wasn't really a cop. He wasn't even a man.

He was the Creator of Heaven and Earth. He was the Rock of Ages, the Light of the World. He was the divine being who'd shown me the Theory of Everything.

He was God.

And yet he looked like an ordinary middle-aged New Yorker, an overweight, overworked guy from Flatbush or Canarsie or one of the other Brooklyn neighborhoods. He was like someone you might see eating a hot dog and carrying a beach chair on the Coney Island boardwalk. Or like one of those scruffy, noisy baseball fans who sit in the cheap seats at Yankee Stadium and bellow, "You suck!" at the home-plate umpire.

The air conditioning came on. The driver could hear my thoughts.

"So how are you doing, Joan? You've been busy since the last time we talked, huh?"

The question was infuriatingly casual. Given the agonies I'd endured over the past 48 hours, his nonchalance was almost insulting. This God was so different from what I'd expected, so unlike all the descriptions in the Bible and all the paintings and sculptures in church. The sight of him didn't fill me with wonder. I didn't sense his overpowering goodness or his all-encompassing love. He was too normal, too human. Not awe-inspiring at all. I was more annoyed than awed.

My disappointment was so intense that I started to wonder again if the whole thing was a hallucination. *This can't be right. Maybe I went over the edge when the cop arrested me. Maybe I'm having another breakdown.*

He let out a sigh as he steered to the rightmost lane of Second Avenue. Then he made a right turn onto Houston Street, which had practically no traffic.

"I don't blame you for being annoyed. Or confused. I get it, it's disturbing. To see God up close? In the flesh, so to speak?" He looked at me in the rear-view. "It throws you off balance, right?"

That was the understatement of the millennium. But I kept my mouth shut and looked out the backseat window. I stared at the dark streets and shuttered stores while he talked.

"I gotta give you credit, though. Even with all the confusion, you've done a great job so far. You noticed the clue I gave you last time, the picture of the extra dimensions, and that wasn't easy to spot. I mean, I knew you were smart, but I wasn't sure how quickly you'd figure out the theory."

I suppose this was meant to be a compliment, but it just made me angrier. God was playing games with me. He was testing me, evaluating my skills, making me run through some kind of theoretical obstacle course. And I didn't appreciate it.

I turned back to the steel grate between us and glared at the rearview mirror on the other side. "Why did you even bother with clues? If you wanted me to know the Theory of Everything, why didn't you just show me the equations right from the start?"

He nodded. His eyes moved up and down in the mirror. "Yeah, good question. The short answer is, I couldn't have done it any other way. I have to operate under certain restrictions when I'm in your world."

"Restrictions?" I leaned closer to the grate. "What are you talking about? If you're God, who can restrict you?"

"You're making a faulty assumption. You think I'm omnipotent, right? The all-powerful, unconquerable, invincible Almighty?" His voice took on a humorous tone. He was making fun of himself. "But the universe isn't set up that way. I can't go around doing anything I want."

"Why not?"

"Think about it for a second. Try to imagine a world where I could do anything I pleased. You know, give food to all the starving people, cure all the cancer patients, stop all the murderers from murdering. What do you think would happen?"

"Sounds like a pretty good world, actually. A big improvement over the status quo."

"But everyone would see what I was doing. The human race would realize that I was in charge and that I would stop all the bad things from happening. In that kind of world, people would have no freedom to make moral choices, because I would prevent them from choosing anything bad. No one would be truly independent or free."

"So what?" I was still annoyed. And when I'm in a bad mood, I'll argue with anyone, even God. "If giving us freedom leads to murder and child abuse and genocide, then maybe you gave us too much. Ever think of that?"

He slowed the car as we approached the intersection of Houston Street and Sixth Avenue. "Without freedom, all people would just be extensions of my will. And my goal was to create a reality that was separate from myself. That was the whole point behind the creation of this universe." The traffic light at the intersection turned red, and he brought the car to a stop. "I can't do anything that would interfere with your choices. I can't reveal my presence in this world, or alter its physical laws in any way that would be detectable. Basically, I have to stay invisible. When you think about it, it's the very opposite of omnipotence." He took his right hand off the steering wheel and pointed at the traffic light ahead. "See? I even have to stop for the red lights."

I frowned. I didn't buy it. "You're not invisible now. At least not to me."

He didn't respond right away. He stalled for a while by turning on the car's windshield washer, which squirted cleaning fluid all over the glass. Then, as the wiper blades thwacked back and forth, he looked over his shoulder at me. "I came to you because we're facing a crisis, an imminent threat. If we do nothing, five billion people will be dead in two weeks."

He stared at me through the grate for a few more seconds, letting the bad news sink in. Then the traffic light turned green, and he faced the clean windshield. In silence, the car rolled forward, still heading west on Houston Street.

I was scared. No doubt about it. When God tells you that a catastrophe is coming, it's a good idea to believe him. I grabbed the armrest on the car's rear door, trying to steady myself. "What kind of crisis?"

He shook his head. "Sorry, I can't reveal the details of—"

"Is it war? A nuclear war?" I shivered in the backseat, which was now cold from the air conditioning. "What else could kill so many people?"

"Look, I'm already bending the non-interference rules to the breaking point. I can't tell you much more. But I *can* say that you've already seen a glimpse of what's coming." He lifted his hand off the steering wheel again and pointed his thumb behind him. "Do you remember what the Russian talked about, back at the diner? The computer program that Andrei's father wrote?"

I didn't want to think about Vlad. I was still traumatized, I guess, still shocked that he'd hit me. But aside from the pain and humiliation, what I remembered most was my bewilderment. The computer program was the reason why Vlad had confronted us. He'd seemed obsessed with keeping it secret, which was why he'd tried to steal our math notebooks. But he'd never explained why the software was so important.

My stomach lurched. Now I was scared and nauseous. "Yeah, I remember. So that program is connected to the crisis?"

He didn't answer, but when I looked in the rearview mirror I thought I saw him nod. That made the nausea even worse. "But what's the connection? What does the program do?"

"All I can tell you is that it's related to the equations in your notebooks. What you call the Theory of Everything." He slowed the car again as we approached the end of Houston Street, at the western edge of Manhattan. Then he made a right turn onto West Street, the broad boulevard that runs alongside the Hudson River. "You know, it's a good thing you retrieved your notebooks from that Russian. From now on, you'll need to keep them very close."

"Why?"

"In the wrong hands, the theory could be dangerous. When unscrupulous people learn something new, something that expands their knowledge of the universe, their first impulse is to misuse it." He stepped on the gas, and the police car sped uptown. We zipped past Manhattan's darkened skyline on the right and the Hudson River piers on the left. "Look what happened after Faraday discovered the principles of electromagnetism. And after Einstein and Bohr revealed the nature of the atom. The human race used its new knowledge to develop better tools and machines. But they also built more powerful weapons. Like missiles and atomic bombs."

I looked down at the notebooks in my hand. I was already clutching them pretty hard, but I tightened my grip a little more. "And the same thing could happen if the wrong people saw the Theory of Everything?"

He nodded. "It would be especially risky if a spy agency or military organization got hold of the equations. It doesn't really matter if the government agency is Russian or American or Chinese, because all of them would do the same thing. They'd use the fundamental laws of nature to build weapons of annihilation."

My head was spinning. The more he told me, the more panicky I got. "Okay, now I'm totally lost. If the theory is so dangerous, why did you show it to me?"

"Because the revelation of the theory can also have positive effects. If it's handled properly." The eyes in the rearview mirror locked with mine. "You need to share the theory with peacemakers, not warmongers. You have to reveal it to a group of international authorities who will spread the news in a responsible way." His voice rose. It rang with divine passion. "When people around the world see the theory in all its glory—the eternal design of the universe!—it'll trigger a change in attitudes, a shift in the human outlook. And all kinds of blessings will flow from that change. It'll defuse the imminent crisis and save billions of lives. It might even end the threat of war, permanently."

"Uh, excuse me?" I peered at him through the grate. "I probably shouldn't say anything, because you're the all-knowing one, right? But don't you think you're being a little unrealistic?"

"I understand. You're skeptical. You assume that people will never stop fighting each other. And that peace is a hopeless dream." He shook his head. "Those statements have been true for thousands of years. But now everything's going to change."

"Because of a theory? A bunch of equations that most people won't understand?"

"Yes. The truth will set you free. Free of hatred and fear." He punctuated this last sentence by rapping his knuckles on the steering wheel. "That's your task, Joan. Revealing the truth. That's the mission I'm giving you."

I turned away from him. I didn't want to hear this. I'd barely started believing in this God, and already he was putting me to work. He'd just given me the hardest job in the history of divine missions. He was treating me like Saint Paul, the apostle who got his head chopped off.

Instead of responding, I looked out the backseat window again. The car was approaching Midtown now, cruising past the helipad on West 30th Street and the Javits convention center on 34th. There were no late-night partyers in this neighborhood, but I spotted a few homeless people trudging down the sidewalk. Up ahead I glimpsed the silhouette of the Intrepid, the aircraft carrier that had been turned into a military museum, docked at one of the Hudson piers. I usually paid no attention to it when I came down this street, but now I stared at the ship as we sped closer. I could see the dark outlines of the warplanes on the carrier's deck, the vintage aircraft from World War II and the helicopters and jets from more recent wars.

It wasn't a comforting sight.

I turned back to Captain Carpenter, searching for his eyes in the rearview. "I'm not the right person for this mission. If you want to share the theory with the world, you should get a famous scientist or mathematician to do it. Someone who's good at explaining complicated things."

He shrugged. "You seem pretty good at it. You explained the theory to Andrei."

"And there's a bigger problem. I'm seventeen. No matter what I say, no one's going to take me seriously."

"Or maybe they'll take you even *more* seriously, because they'll be so shocked and amazed at how much mathematics you know."

This was frustrating. I started to wonder if it was even possible to win an argument with God. "Okay, here's the biggest problem—I don't know any 'international authorities.' What am I supposed to do, call the United Nations? Tell them about the Theory of Everything over the phone?" I frowned as I tried to picture it. "See, it would never work. That's why you need someone famous for this job. Like maybe a Nobel Prize winner."

"No, that won't be necessary. There's an easy way for you to contact everyone you need to talk to." His voice was lower now and very calm. "You can do it this weekend. On Sunday."

I was so surprised, I pulled away from the grate. I leaned back against my seat, recoiling from him. "Sunday? *This* Sunday?"

"There's a conference this weekend in Albuquerque, New Mexico. It's a meeting of five hundred physicists, mathematicians, philosophers, and theologians. The conference's title is 'Bridging the Gap Between Science and Religion,' and they've invited some of the world's smartest thinkers. Many string theorists will be there, including Edward Witten. And spiritual leaders too, like the Dalai Lama." He smiled in the rearview mirror. His NYPD face softened and became almost beautiful. "On Sunday morning they're holding a panel discussion called 'The Search for New Ideas.' It's the perfect opportunity for you to unveil the solution to string theory."

I shook my head. I couldn't believe it. God had planned the whole thing for me. He had it all figured out.

In a daze, I stared at the back of his neck. "So did you buy the plane tickets already? Or were you planning to teleport me to New Mexico?"

"I'm afraid you'll have to travel by car." He stopped smiling. His hard, serious police face came back. "You can't fly there. Your parents wouldn't allow you to go, and you don't have your own credit card to buy a ticket."

"Uh, I don't have a car either. And I don't know how to drive. I take the subway everywhere."

He made a dismissive gesture, flicking his wrist, brushing away all the difficulties. "Don't worry. God will provide."

By this point, he'd steered the patrol car onto the elevated highway that ran between the Hudson and Riverside Park. We were going at least sixty miles per hour, and in less than a minute we got off the highway at the 79th Street exit. Now we were only two minutes away from my apartment building, and I was very grateful to be close to home. I wanted to get in bed and sleep for a hundred hours. I wanted to forget that this night ever happened.

But as he stopped for another red light at 79th Street and Broadway, he looked at me over his shoulder again. "I'm sorry, but you can't rest for long. The Russian knows your address, and he's going to come after you. I can't force you to go to New Mexico, but for your own safety you'll need to leave New York very soon."

My throat tightened. I gazed in terror at the familiar streets of my neighborhood, the subway station, the bus shelters, the darkened Chase Bank branch. At any moment I expected to see Vlad slink around the next corner. "You think he'll come in the next few hours? Before morning?"

"No. But you should get out of the city by the afternoon. Definitely before it gets dark."

"What about my parents? Will he come after them too?"

"Just make sure you take your math notebooks when you leave. He'll probably put your apartment building under surveillance by this afternoon, and if he sees you leave the building with the notebooks, he'll know there's nothing for him here. Then your parents will be safe."

The traffic light turned green. He made a right turn on Broadway, then a left on 78th Street, and drove the last few hundred feet to my building. He stopped the car in front of the entrance, and I leaned toward the backseat window so I could look up at the fourth floor. To my relief, I saw no lights in the windows. Mom and Dad were still asleep. They hadn't noticed that I'd slipped out.

I'll write them a note before I leave the city. I'll invent a believable explanation and promise to come back in a couple of days.

Oh crap, who am I kidding? Leaving a note won't help. Mom and Dad will go ballistic as soon as they read it. They'll go out of their minds with worry.

Captain Carpenter stepped out of the patrol car and opened the rear door for me. The middle-of-the-night darkness lay heavy on 78th Street. I felt a terrible dread as I got out of the backseat, with both of the math notebooks under my arm.

But the captain gave me a parting gift: he smiled again. It was such a simple thing, and yet it was so dazzling. For a split second he lifted the darkness. The buildings on the street seemed to shine under his gaze.

"Be of good cheer, Joan. I know you can do this."

I bit my lip. I wanted to believe him, but it was hard. "Are you sure you can't find someone else?"

"Samantha was right about you." His smile grew even brighter when he said my sister's name. "You underestimate yourself."

"What?" All at once I felt dizzy. "You…you talked to her too?"

"I heard her thoughts. She was very proud of you." He cocked his head, still smiling. "And so am I."

This was too much, too painful. I had to turn away. I went to my building's front door and fumbled in my pocket for the key. My hands were shaking.

Okay. I get it. Sammy was a good person. She loved the world and everyone in it. She never passed up a chance to help people. And she always encouraged me to do the same.

So I'll go to New Mexico. I'll do it for her.

I finally got the key into the lock and opened the door. As I stepped inside, I looked over my shoulder.

Captain Carpenter and the police car were gone.

Chapter Seventeen

A phone call woke me up eight hours later, almost noon on Thursday. I stretched my arm out from under the bedcovers, blindly grabbed the buzzing iPhone from my night table, and held it in front of my bleary eyes to see who was calling.

It was Charlotte.

I bolted upright in bed, instantly frantic. I didn't want her to know she'd awakened me, so I coughed and cleared my throat to make sure I didn't sound groggy. It was just a phone call, and I knew I was getting way too excited and terrified, but I couldn't help it. *It's Charlotte! She wants to talk!*

I answered the phone, straining to be as peppy as possible. "Hello?"

"Holy crap, Joan, are you all right?" Her voice was high-pitched and agitated. "Where are you?"

"Uh, I'm home. In my bedroom. And I'm fine. What—"

"Oh, thank God! I thought you were in jail or something. The last thing I saw on the video was the cop pushing you into the patrol car."

I raised a hand to my forehead. *Oh no. Not again.* "A video? Another one?"

"Yeah, on YouTube. You haven't seen it yet?" Charlotte took a deep breath, and her voice calmed a little. "It's insane. It's getting even more views than your last one. You're running down the street with a flag, chasing some weirdo in a long coat? And you're shouting about God?"

I thought back to the night before. I remembered sprinting past the crowd of late-night partyers on Second Avenue and seeing them point their phones at me. I guess it was predictable that at least one of them would put the video online. How could they resist?

"Hey, Joan? You still there?"

I cringed as I thought of Charlotte watching me on YouTube. It was mortifying. "Yeah, it's all true. I went down to the East Village last night and things got a little crazy."

"Girl, why didn't you take me with you? If you'd called me up, I would've gone down there in a heartbeat! Do you know how bored I was last night? I was stuck in this hotel room with evil Teresa, listening to her complain about the room service."

Despite the embarrassment, my heart leapt. Charlotte had just said she wished I'd invited her to the Village. I thought of the couples I saw last night in front of the dance club, the pairs of women holding hands. "Sorry, I should've told you about it. The thing is, I went downtown for a study session with someone from my school, and I thought it would be boring. Then all the craziness happened."

"Oh, it was more than just crazy. It was epic! What you did in that video was *amazing*." Her voice fizzed with enthusiasm. "When you pointed your flag at that weirdo? And yelled 'This is the will of God'? I mean, I didn't understand who you were shouting at or why, but it was freakin' awesome."

I liked hearing this. It lessened my distress. Why should I be embarrassed if Charlotte thought I was awesome? Why should I care about anyone's opinion but hers? "So how did you hear about the video? Please don't tell me it's trending already."

"I got a message from Marguerite, my best friend. I told her about you yesterday and sent her the link to your last video, the one in the subway station. When she went on YouTube this morning she saw the new video and recognized your face."

My emotions whipsawed. This was the first time Charlotte had mentioned her "best friend." I tightened my grip on the phone. "Marguerite? She's, uh, a friend from home?"

"Actually, she's from Pennsylvania. She's one of the Everlights."

"Everlights?"

"I told you about them, remember? They're the people in the chat group on Everlasting-Light-dot-com, the website of my dad's church. I've never met Marguerite in person, but we talk online almost everyday. So we're pretty close."

Now I didn't know what to think. Should I be relieved that Marguerite was just an online friend? Or should I be worried about the "pretty close" remark? I had to stop thinking about it, I was getting too obsessed. "Can we get back to the video? You said your friend recognized me?"

"Yeah, go on YouTube right now. You gotta see what she wrote in the Comments section. It's seriously righteous."

"Okay, give me a minute."

I reached for my laptop and went on YouTube, heading straight for its list of Trending videos. Sure enough, at the top of the list was "Crusader on Second Avenue," which had been posted seven hours ago by someone named Duke Riverbend. It had already been viewed three million times.

I clicked on the video but immediately paused it. I had no interest in reliving last night's insanity. Instead, I scrolled down to the Comments section and quickly found a very popular comment—it had two thousand like's—submitted an hour ago by "Sister Marguerite of Alverton":

Oh my brothers and sisters! The Holy Girl who fights for THE LORD in this video is the same Blessed Messenger who shared her vision of GOD with us a few days ago. (Go to this YouTube link for "Crazy God Girl in NYC Subway Station" and see for yourself!) What's more, she's a cousin of my good friend Sister Charlotte, spiritual disciple of the Church of Everlasting Light, who has informed me that the girl's first name is Joan. She's being persecuted by Godless psychiatrists trying to convince her that she's mentally ill and that her heavenly messages from THE LORD are nothing but illusions. BUT GOD WILL NOT BE MOCKED!

Below Marguerite's comment were 152 replies. I scrolled through the most recent ones, which included the usual mix of affirmations and insults:

Sister Emma Dowling, 1 minute ago

Thanks for the revelation, Marguerite! Joan seems truly special. May God bless her forever. (P.S., I'm an Everlight too!)

Shoobie Doobie Man, 3 minutes ago

What a crock. The girl is obviously ill. You're not doing her any favors by encouraging her.

John Paul Hedgehog, 4 minutes ago

How old is she, Marguerite? I wouldn't mind meeting her. I like the crazy ones.

Lily Chang of the Valley, 6 minutes ago

Funny coincidence, isn't it? A girl named Joan who gets messages from God? Is she French?

Hot Mess the First, 7 minutes ago

Welcome to NYC, where there's a nutcase on every corner.

Star Phoenix, 8 minutes ago

We should call her Saint Joan of New York.

I'd lowered the iPhone from my ear, but as I stared at the laptop I heard Charlotte's voice, tinny and distant, come out of the phone's speaker.

"Joan? Did you find it?"

I felt drained, demoralized. I wanted to forget the whole thing and go back to sleep. But instead I raised the phone. I did my best to put the pep back into my voice. "Yeah, I'm reading the comments. I guess I should look on the bright side. I got everyone's attention."

"Listen, who was the guy in the long coat? And why were you chasing him?"

I shook my head, even though Charlotte couldn't see me. I couldn't answer her questions now, not over the phone. "It's a long story. I want to tell you about it, but we need to talk in person."

"Well, let's get together. How about this afternoon? You took off from school today, right?"

As I considered the idea, I started to feel a little better. Yes, I still had to leave New York by nightfall. And yes, I had to go on a cross-country mission I didn't ask for, and make a pilgrimage to a Science and Religion conference in New Mexico that I didn't really understand. But if I could see Charlotte again before I left, at least it would give me a smidgeon of consolation. Maybe that would be enough to get me through the journey.

"Yeah, sure, I can get together. I have that appointment I told you about, with the psychiatrist my parents found, but that's at one o'clock, so it should be finished by two. Can we meet then?"

"Two o'clock would work. We can even go for a drive if you want. Teresa rented a Chevy Tahoe from the Avis rent-a-car place, but she doesn't plan to use it till tonight, so the SUV is just sitting in the hotel's garage right now. And I'm in the mood for a little joyride."

This piece of information was so surprising, I got up from my bed. I stood in the middle of the bedroom, holding the phone in my trembling hand. "You know how to drive? You have a license?"

Charlotte laughed her beautiful pattering laugh. "Of course I have a license. For your information, nearly every seventeen-year-old in America knows how to drive. It's only you New York City kids who don't learn, because you love your subway so much."

"And you can just go down to the hotel's garage and take out the car?"

"Yeah, no problem. Teresa went out to do some shopping, but she left the valet ticket for the Tahoe here. And it would serve her right if I took the car away from her." She paused. When she spoke again, her voice was a lot less cheerful. "To tell you the truth, I'd like to push her SUV into the river. You know where my stepmother's planning to go tonight?"

In just the past few seconds, Charlotte's tone had darkened. She sounded so angry now, I was scared to ask the next question. "No, where?"

"She's going to leave me in the hotel and meet her boyfriend in New Jersey. He's a rich arrogant jerk who has a beach house on the Jersey Shore."

"Wait, how do you know this?"

"Teresa's pretty bad at covering her tracks. Ever since Dad got sick, she's been fooling around with that jerk. What makes it even worse is that he used to be one of Dad's business partners."

"Oh man. That's gross."

"He comes to New Mexico once a month, supposedly for business reasons, and Teresa meets him at his hotel. In fact, that's why she hasn't tried very hard to get Dad out of the New Life Oasis center. It's a lot easier for her to visit her boyfriend when her husband's cooped up in the loony bin."

I nodded. Now I understood why Charlotte was so angry. "I'm sorry. That really sucks."

"I hate her so much. I just wish I could get away from her." She was near tears, I could tell. "As far away as possible."

If we were having this conversation in person, I would've put my arms around Charlotte. I would've hugged her tight and kissed her cheek. But all I could do over the phone was say "I'm sorry" again. "Is there any way I can help?"

"No, there's—"

"Maybe I can help you push Teresa's rental car into the river? Or slash its tires?"

She laughed. At least I'd made her laugh. "If I were smart, you know what I'd do? I'd hop into that SUV and drive all the way back to Santa Fe. And then I'd go straight to the New Life center and get Dad out of that nuthouse. I'd storm into the place and just grab him, you know? And then we'd go live in the wilderness somewhere."

Her voice cracked. She didn't allow herself to cry, but she sounded so defeated. She clearly wasn't serious about storming the New Life center. She was just thinking out loud and letting off steam. Her heart was in the right place—she wanted to rescue her dad—but she didn't have the conviction to pull it off. She wasn't desperate enough to steal the rental car and drive it across the country.

But I was.

"Charlotte, we can do this together. We can save your dad. We just need to make a plan, okay?"

God had said he would provide. And he'd kept his promise.

* * *

Ten minutes after I got off the phone with Charlotte, Dad knocked on my bedroom door. He said we had to leave in half an hour for the psychiatrist's office, so I went to the bathroom and got ready. I checked my face in the mirror and saw no bruise where Vlad had hit me, just a bit of redness. I covered it with a dab of Mom's Maybelline foundation, then made myself a quick lunch, a turkey and cheddar sandwich. Then I returned to my bedroom and started collecting everything I would need.

I opened the underwear drawer in my bureau and extracted all my cash from its hiding spot inside a tube sock. I'd earned $400 over the summer from a tutoring job at the Manhattan Math Academy, but I'd spent most of it on the two pairs of running shoes I'd bought for the cross-country season. Only $138 was left. After doing a few rough calculations in my head—inputting the distance to New Mexico, the typical fuel economy of an SUV, and the price of gasoline—I realized I'd need at least $300 just for the gas. But I couldn't worry about that right now.

I slipped the cash into the right pocket of my jeans. I put my iPhone in the left pocket, and in each of the back pockets I stuffed an extra pair of underwear. I had to travel light because I couldn't carry a bag. If I took a duffel or even a backpack to the psychiatrist's office, my parents would definitely look at me funny. Even worse, it would raise the suspicions of anyone else who might be watching me. Assuming that Vlad had actually put my apartment building under surveillance, he'd be on the lookout for signs that I was planning to leave the city. If he saw me carrying a bag out of the building, he might swoop in and try to stop me from bolting.

Last but not least, I picked up the two math notebooks—mine and Andrei's—and a couple of pencils. I gave my bedroom a final once-over, glancing at the unmade bed, the beanbag chair, the stacks of textbooks on my desk, and the bras and T-shirts scattered across the floor. I was leaving the place a mess, and I hadn't written a note for my parents either. But I'd figured out a different way to say goodbye to them. It would be less formal and ominous than a written note, less likely to drive them crazy with worry. At least that was my hope.

I left the room and went to the foyer, where Mom and Dad were already waiting, impatient to leave. Mom frowned when she saw how I was dressed. "Well, that's an interesting message on your shirt. Is that going to be your attitude toward the psychiatrist?"

I looked down at my T-shirt. The word RESIST was written in big black letters across my chest.

I shook my head. "No, it's random. I had only two clean shirts in my drawer, and the other one said, 'ALL PSYCHIATRISTS ARE GREEDY QUACKS.'"

Dad chuckled. "You made the right choice then." He pointed at the notebooks in my hand. "What about those? You're planning to take notes?"

I told him the truth. "There's something I want to show the doctor."

Mom and Dad cocked their heads, almost simultaneously. They were obviously curious, but they didn't pry. They had great respect for the rules of psychotherapy, mostly because they'd both gone to therapists at various points in their lives and seen some benefit from it. And they firmly believed it would help me too. They had faith in the process.

Unfortunately, I didn't share their faith. But I smiled at them anyway, hating myself for the deception. "All right, I'm ready. Let's get this over with."

We took the elevator down to the lobby. The psychiatrist's office was on Central Park West and 82nd Street, just a few blocks away, so the plan was to walk there. My parents flanked me as we stepped out of our building, Mom taking position on my left and Dad keeping to my right. They must've thought they were being protective, but it was actually kind of creepy. I felt like a prisoner being escorted down Death Row.

Dad pointed at the trees on our block. "Look at that. The leaves are starting to change color."

Mom nodded. "Yes, it's beautiful. I love this time of year."

I didn't look at the foliage. Instead, I scanned the sidewalks on both sides of 78th Street, searching for suspicious characters. I wasn't sure if I'd see Vlad himself—after the incident last night, he probably wouldn't be so brazen—but I thought I might spot an accomplice or two. If Vlad were truly serious about this operation, wouldn't he recruit someone from the Russian embassy to help him? Maybe even a whole team of operatives to keep watch over me?

and then he won't bother with punishment. He'll just kill us. He's done it many times before."

We walked on in silence, turning right at 82nd Street and crossing Amsterdam Avenue. I looked over my shoulder at Mom and Dad, who remained about ten feet behind us, and they looked back at me with evident concern. Andrei swayed a little as he walked, struggling to stay upright, and I wondered if he'd gotten any sleep the night before. I tugged his arm, trying to snap him out of his daze. "Tell me about the computer program that your dad worked on. Why is Vlad so worried about it?"

Andrei sighed. "That's something else I feel guilty about."

"Why?"

"Do you remember what I told Vlad at the diner? How I insisted there was no connection between my father's program and the theoretical physics we were working on?"

"Sure, I remember."

"That was a lie. My father's specialty was developing programs for quantum computers. Which are very much related to the fundamental laws of physics."

I knew what a quantum computer was. The teacher in my Advanced Placement physics class had spent a whole week talking about the technology. Basically, a quantum computer takes advantage of the laws of atomic physics to radically increase its computing power. The machine isolates individual atoms and puts them into "superposition," a weird state in which an atom can do two opposite things at the same time—for instance, its magnetic orientation can be up *and* down simultaneously. Then the computer manipulates the atoms using laser beams and magnetic fields. It arranges the particles like the beads on a microscopic abacus, enabling them to perform trillions of calculations in a fraction of a second.

To be honest, I thought the idea sounded like science fiction, but the physics teacher assured us that thousands of researchers around the world were racing to build the first quantum computers. Their main motivation was the fact that the new machines could perform one crucial task very well, a billion times faster than any ordinary computer could.

The task was breaking secret codes.

I tugged Andrei's arm again, harder this time. "Okay, now it makes sense! It's a code-breaking program, right?"

He seemed half-asleep, his eyelids drooping. But he nodded. "In the last two years, my father worked night and day on the project. Vlad's agency had promised him three hundred million rubles if he could write a successful program. But when he was almost finished with it, he discovered what the agency planned to do with the software."

"What was their plan? To break all the American codes? So the Russians could launch a sneak attack somewhere?"

Andrei shrugged. "I don't know. But whatever it was, it destroyed my father. He deleted his computer files and burned his papers, erasing all his work on the software. Then he ran across St. Petersburg to the Obukhovsky Bridge and threw himself into the Neva River."

Another silence fell. I squeezed Andrei's arm but he didn't respond.

For the next two minutes we walked down the long block between Amsterdam and Columbus avenues. We trudged past brownstones and red-brick apartment buildings, past a beautiful church and an ugly police station. Andrei moved like a zombie, his brown shoes clomping on the sidewalk, his blank face bobbing. And I felt so sorry for him that I did something I would regret later on. I squeezed his arm again and stretched my other hand to his face so I could caress his cold cheek. At that moment I didn't care if Andrei got the wrong idea about my feelings toward him. I also didn't care if my parents saw it and got the same wrong idea. The boy needed some kindness. He was in a stupor.

"You don't deserve this either," I whispered.

He shivered at my touch. When he looked at me again, his eyes were wet. "No, I'm guilty. At least partially. My father taught me how to write the programming language he was using. He knew I liked puzzles, and this was a tricky one, figuring out how to program a completely new kind of computer. We started working on it together."

"So you wrote some of the code-breaking software?"

"Yes, a large part of it. After my father's suicide, Vlad was determined to reconstruct the program, and he ordered me to tell him what I knew, all the lines of software code I could remember. We made a deal—I would give him what he wanted if he let my mother and me leave Russia." He shook his head. "But Vlad never trusted me. He's been following me and watching me ever since we came to New York. He thinks I'm going to tell the American agents about the software and ruin whatever operation he's planning."

I didn't like the sound of that. I got the sense that the Russians had already put a catastrophic strategy in motion. My legs felt heavy as we crossed Columbus Avenue, and my stomach clenched. I thought of the prediction God had shared with me in the police car last night. *If we do nothing, five billion people will be dead in two weeks.*

"Andrei? We can't let this happen. We need to fix this."

"I'm sorry, but I can see only one solution." He winced. "I should do what my father did."

"*No!*" I stopped in my tracks and yanked his arm. "Don't talk like that!"

My parents stopped too, and it looked like Dad was about to say something. But I quickly faced forward and resumed walking, pulling Andrei down the block. "Listen carefully, all right? The first thing we need to do is get out of New York. We need to get away from everyone who's watching us, understand?"

Andrei nodded, but he looked perplexed. "How will we—"

"Just listen." I pointed ahead at the tall apartment building on the corner of 82nd Street and Central Park West. It was less than a hundred yards away. "My appointment is at a psychiatrist's office on the 14th floor of that building. The doctor sees patients in his apartment, and that's a lucky break for us."

"Lucky? Why?"

"It's a fancy building for rich people, which means it has big, luxurious apartments. And a big, fancy apartment usually has a service entrance near its kitchen. You know, to make it easier to haul the trash to the building's freight elevator."

Andrei's perplexity only seemed to deepen. "And how does that help us?"

"I already looked at the building on Google Street View." I pointed at its ground floor. "The main entrance is on Central Park West. That'll probably be the focus of the Russian agents. They'll watch us go into the building for the appointment, and they'll keep an eye on that entrance while they wait for us to come out. But the building also has a parking garage. Its entrance is around the corner on the 82nd Street side. See it?"

Andrei looked ahead, peering at the garage's driveway, which sloped downward from the street to the building's basement. Then he turned back to me and raised his eyebrows. "You think it's possible to take the freight elevator down to the garage?"

I nodded. "Here's the deal. You're going to come with us to the doctor's apartment. I'll tell my parents that I need you there for moral support. And while I'm talking to the psychiatrist, you're gonna do a little recon. You know what that means?"

"Recon? You mean reconnaissance?"

"Exactly." I gave him a smile. "You're going to scope out our escape route."

Chapter Eighteen

The psychiatrist's apartment was just as big and luxurious as I expected. The reception area had a Persian rug, a crystal chandelier, and a stunning young receptionist who greeted us at the door. There were arched entryways to the left and right, and through the one on the left I glimpsed the doctor's living quarters, specifically a glamorous dining room with antique chairs and a long gleaming table. But the receptionist escorted us through the doorway on the right, which led to a waiting room with a more professional décor—plain upholstered chairs, copies of *National Geographic*, a couple of tasteful paintings on the walls.

We didn't have to wait long. There weren't any forms to fill out, because Mom and Dad had already consulted with the doctor and given him my medical history. While my parents settled into their chairs and Andrei went looking for a bathroom, the receptionist opened a door on the far side of the room and beckoned me toward her. "Come inside, Joan," she said. "Dr. Cauchon is ready to see you."

My teeth started to chatter. Although everything was proceeding just as I thought it would, I still felt a strong, quaking dread. I took a deep breath as I stepped toward the door, holding the math notebooks under my arm. Then I marched into the psychiatrist's office.

What I noticed first were the books. Thousands and thousands of them were stacked on the shelves of the floor-to-ceiling bookcases, which covered three of the room's walls. Most of the books were thick tomes with dull gray covers and gold lettering on their spines. They were probably medical encyclopedias, manuals that listed every psychiatric disorder in the universe.

It was total overkill. There were so many books on the shelves that they deadened the sound in the room and gave off a musty odor. And because the office had no windows, it seemed weirdly dark. The only light came from a squat lamp on a fancy wooden desk that had nothing else on it—no phone, no computer, not even a single pen or pencil. The desk's bare polished surface glowed under the yellow lamplight, which also illuminated the face of the man sitting behind it.

Dr. Cauchon was old, at least seventy. He wore a dark blue suit and a striped tie and a pair of gold cufflinks. He was almost bald, with just a fringe of gray hair above his ears, and his nose was dimpled at the tip. The most alarming thing about him, though, was the mole on his forehead. It was large, more than half an inch across, and it hung just above his straggly left eyebrow. But what made it so disturbing was its color. It was deep red, like a setting sun.

He studied me as I approached his desk. Wrinkles furrowed the skin around his eyes and mouth. "It's a pleasure to meet you, Miss Cooper." He pointed at a black leather chair in front of his desk. "Please, sit down."

His voice was soft and calm, which I supposed was a useful trait for a psychiatrist. I sat in the chair and rested the notebooks in my lap. It was difficult to hide my dread.

Cauchon smiled, trying to put me at ease. "I've been looking forward to this meeting. Your parents have already told me so many things about you, including your formidable talent at mathematics. It's very impressive."

He waited for a thank-you, but I was in no condition to be congenial with this man. The only way for me to survive this appointment was to be business-like and efficient, as serious as possible.

"When you talked to my parents, did they also tell you what I did at the 125th Street subway station?"

Cauchon lifted his left eyebrow. The straggly gray hairs edged closer to his mole. "You don't waste any time, do you?"

"Well, I know we only have an hour, and I have a lot to say."

He nodded. "Then let's get right to it. Yes, your parents and I talked about that episode. But now I'm hoping to hear about it directly from you." He leaned forward in his chair. "My understanding is that when you woke up at Bellevue Hospital, you didn't remember anything that happened on the sub-way platform. I'm wondering, though, if any memories have come back to you since then."

"No, none at all. But I've experienced another, uh, episode. I remember this one very clearly, and now I know what's going on."

I'd stopped shivering. My strategy was successful: once I started talking, my panic receded. And Dr. Cauchon seemed interested in hearing the story. He leaned a bit farther across his desk, and the lamplight shone on his dimpled nose. "When did this second incident occur?"

"Last night. You'll be skeptical when I tell you this, but you need to hear me out." I took another deep breath. "I've talked with God. On three separate occasions. And God has given me a mission."

To my relief, Cauchon didn't smile or chuckle. He didn't even lift his eye-brow. He simply nodded again, as if I'd told him that I was failing in school or fighting with my parents or some other perfectly ordinary problem. "What is this mission?"

His calmness flustered me. The conversation was progressing much more smoothly than I'd anticipated. I reached for my math notebook and fumbled through its pages until I found the pencil sketch of the Calabi-Yau manifold. "I could give you a crash course on string theory, but that's not really neces-sary, okay? All you need to know is that this diagram is the key. God showed me this picture and helped me figure out the fundamental equations of the universe. And now God wants me to spread the news to the whole world."

Cauchon nodded a third time. "Very interesting. I'm beginning to under-stand the videos now."

My throat tightened. "Videos?"

"Yes, I found two of them. On YouTube." He reached into the drawer of his desk and pulled out an iPad. "After I consulted with your parents, I wanted to learn more about what happened to you at the subway station. So I took the liberty of doing an Internet search." He held the iPad in the air for a moment, then slipped it back into the desk drawer. "When I searched for your name and the name of your high school, I found a comment from a classmate of yours who recognized you in the first video. And just an hour ago I did another search and found the second one."

I was floored. I hadn't thought that psychiatrists would be big fans of YouTube. Which was stupid of me, I guess; it was probably their favorite website. How could they resist it, with all that clinical insanity on display?

But at least now I knew why Cauchon hadn't been surprised by my confession of faith. He'd already seen me ranting about God on the subway platform and on Second Avenue. "Wait a second. Have you told my parents about the videos?"

"Don't worry, these sessions are confidential. I won't reveal anything you'd rather keep secret, unless concealing it poses a danger to you or someone else."

I didn't believe him. I had the feeling that Cauchon would tell my parents *everything*. But that was okay. My goal, after all, was to use this appointment to send a message to Mom and Dad.

"You know what? You can tell them about the videos. They're not important." I raised my math notebook and pointed at the sketch. "Here's the important thing. String theory is a horrendously difficult problem, and there's no way I could've solved it on my own. That's how I know God exists—She showed me the solution. That's my proof." I tapped the page to emphasize my point. "I can't give you a copy of this sketch, it's too dangerous. But I want you to tell my dad about it. It's called a manifold, a Calabi-Yau manifold. Dad's working on a magazine story about string theory, so he'll understand what I mean about the difficulty of the problem. Maybe this'll convince him that believing in God isn't so crazy."

Cauchon frowned. The wrinkles on his forehead deepened, above and below his mole. "Now, now. There's nothing crazy about religious belief. It brings comfort to billions of people around the world." He kept his eyes on me as he leaned back in his chair. "But sometimes a very strong belief can provoke unreasonable actions. You've studied world history at your high school, so I'm sure you're familiar with some of the extremism inspired by religion?"

I was confused. *Where's he going with this?*

"Like the Crusades? Is that what you mean?"

"Yes, certainly. And the Inquisition. In both of those cases, the believers got carried away. Because they focused so much on their religious faith, they lost sight of ordinary human decency. So they slaughtered their enemies in the name of God. They used religion to justify torture and all kinds of immorality."

I shifted in the leather chair, suddenly uncomfortable. "What are you saying? My belief in God is immoral?"

"The problem isn't your belief. It's your behavior. You caused a scene at the subway station, and it was so disruptive that the paramedics had to take you to Bellevue. This frightened your parents very much. And you apparently caused another scene last night." He raised a veiny hand and pointed at me. "What were you doing with the flag in that video? Were you threatening someone with it?"

I had no intention of discussing this subject. The details about Vlad would only terrify my parents. I wanted to leave them a comforting message, so I wasn't going to tell Cauchon anything about Russian spies or dangerous software.

"Sorry, I can't talk about that. It's a private matter. But I promise you, I didn't do anything immoral or irrational."

"Yes, I'd like to believe you. You seem very rational right now, very calm and composed. In the past few days, though, you've experienced two serious breakdowns. You've suffered memory loss and exhibited abnormal behavior. It's worrisome, to say the least."

I thought back to the emergency-room doctor at Bellevue. He'd also said my behavior was "worrisome." The word was probably in all the psychiatric manuals. "So you think I have schizophrenia? Is that it?"

"It's much too early to make a diagnosis. To be considered a schizophrenic, a patient has to exhibit abnormal symptoms for at least a month. It's possible that a completely different medical problem, such as epilepsy, could've caused this unusual behavior. At this point we can't rule anything out."

"What about the possibility that God actually spoke to me? We shouldn't rule that out either, correct?"

Cauchon frowned again and tilted his head upward. He stared at the thousands of books on the wall, as if looking for inspiration. "Tell me some more about what you experienced during your...your visitations from God. Exactly what did you hear and see?"

I shook my head. "What's the point?" I closed my math notebook and put it back on my lap. "I showed you the proof that God exists, but you don't believe it. You're convinced that it's a delusion and that I'm psychotic. You only want to know the details so you can judge how crazy I am."

His cheek twitched. He seemed irritated. "I'll repeat what I said before. Belief in God isn't the problem. I'm much more concerned about your claim that God has shown you the solution to a mathematical problem. That's not a typical religious experience."

"Really? So religion is acceptable to you only if it falls within the typical boundaries? Otherwise, it's a sign of mental illness?"

"I'm also concerned—"

"And who sets those boundaries? Who decides which beliefs are acceptable? You?"

Cauchon glowered. His face seemed to darken. "I'm also concerned about your so-called 'mission,' and your statement that the drawing in your notebook is 'dangerous.' Those are red flags. They're signs of delusional thinking. And I believe these symptoms may be related to the trauma you've experienced because of the death of your sister. Although I can't make a diagnosis yet, something is clearly wrong."

I picked up the notebooks and rose from my chair. My eyes watered, but not because I was angry at the psychiatrist. I was distressed because it was time to end this appointment. I'd made a plan with Charlotte and Andrei, asking them to meet me at a prearranged time and place. They were waiting for me now, so I couldn't stay the full hour with Dr. Cauchon. I needed to wrap things up and say my goodbyes.

"Okay, I've had enough. This isn't working."

For the first time, Cauchon seemed surprised. He stared at me for a few seconds, his mouth open, his eyebrows arching. Then he glanced at his watch. "But we still have forty minutes left. We haven't even—"

"There's one more thing you can do for me. When you talk to my parents after I go, can you give them a message?"

"What? Why are you—"

"Tell them I won't be gone for long. Just a few days." I had to pause. It was hard to get the words out. A tear slid down the side of my nose. "Tell them I needed a break, so I decided to go on a trip with Andrei. It's a joyride thing, a chance for us to chill for a while. But I promise I'll come home by next week. Can you remember all that?"

His mouth opened wider. He looked thoroughly confused. "Why do you need me to give them a message? Why can't you do it yourself?" He gripped the armrests of his chair and stood up. "Please, sit back down. I'm going to get your parents and bring them in."

I spun around and headed for the door, reaching it before Cauchon stepped past his desk. Then I flung the door open and raced across the waiting room. Out of the corner of my eye I saw Mom and Dad jump out of their chairs, but they didn't run after me. By this point I was red-faced and sobbing. It looked like I was dashing to the bathroom to cry myself out, so nobody followed me.

I charged through the reception area and into Dr. Cauchon's glamorous dining room. Just beyond it was a sparkling kitchen with the latest top-of-the-line appliances. At the far end of the kitchen, Andrei held the service door open for me.

"Come on!" He looked jittery, but he waved me forward anyway. "The freight elevator is here!"

* * *

We took the elevator down to the basement. From there, a corridor led to the building's underground parking garage, where a bright red Tahoe SUV idled in front of a row of parked cars. A garage attendant in a blue uniform stood beside the Tahoe, chatting through the open window with the driver. As Andrei and I came closer, I saw Charlotte behind the steering wheel. She wore a yellow sundress and a pair of lovely feathered earrings.

I wiped away my tears with the back of my hand, trying to clean myself up. Ordinarily, I didn't care much about my looks, and they certainly didn't deserve to be a priority at that moment. But I was glad I did it, because a second later Charlotte turned her head and smiled. She looked so happy to see me. My heart nearly burst.

She twisted around in the driver's seat and stuck her head out the window. "Joan! You're here!" Then she saw Andrei, and her smile faltered a bit. "Who's that?"

"Don't worry, he's a friend." I ran to the passenger side of the SUV and Andrei followed. While I opened the car's front door, he got into the backseat. "Okay, let's move. No time to lose."

Charlotte nodded, then turned back to the garage attendant, a lanky young dude who seemed completely smitten with her. She waved goodbye to him and simultaneously rolled up the driver-side window. "Sorry, we gotta go now."

The window was up before he could ask for her phone number. Charlotte shifted the Tahoe into drive. "What took you so long? I thought you said you'd be in the garage by one-fifteen."

I dropped the math notebooks into my lap and leaned back against the plush seat. "Oh man, don't ask. I got into an argument with the psychiatrist." I pointed over my shoulder at Andrei. "And I'm sorry for springing this on you, but my friend is coming with us. His name is Andrei Mishkin, and he's in a lot of trouble."

Charlotte glanced at the backseat. To my relief, she gave Andrei a smile too, and it was almost as big as the one she'd given me. The change in plans didn't seem to upset her. "What did you do, Andrei? Rob a bank?"

He shook his head, still nervous. "I'm afraid it's more complicated. But before I tell you the story, may I ask a question? Did you notice any unusual people outside the building when you drove into the garage?"

Charlotte cocked her head as she steered the SUV toward the exit. She seemed amused. "Hmm. I don't remember anyone in particular. Are the cops after you? Or just a jealous girlfriend?"

"Uh, neither." Andrei leaned forward and tapped me on the shoulder. "Just to be on the safe side, Joan, you and I should duck down. The agents might be looking at the cars as they come out of the garage."

Charlotte made a left turn, and the Tahoe ascended the ramp that sloped up to the street. "Agents? Wow, that sounds intense. I can't wait to hear your story."

Clearly, she didn't grasp the seriousness of the situation. But I did, so I lowered my head and scrunched into the space below the car's glove compartment. At the same time, Andrei flattened himself into the footwell between the front and rear seats.

Charlotte laughed as she watched us contort ourselves. When she reached the top of the garage's ramp, she slowed the SUV and peered to the right, looking for oncoming traffic on 82nd Street. Then she made another left turn. "Okay, people, the coast is clear. No one's on the sidewalk except that homeless guy over there, and he's busy rooting through the trashcan on the corner. No, wait a minute, now he's crossing the street. Look at that guy, he can barely stand up. I better—*whoa!*"

The Tahoe screeched to a halt. An instant later I saw the homeless man, who'd staggered in front of the car and forced Charlotte to stop. He came over to the passenger side of the SUV, his filthy brown sweatshirt brushing against my window, his hood cinched tight around his face, and I thought he was going to tap on the glass and beg for some change. But then he peered into the SUV and saw me crouched beneath the glove compartment, and he shouted something very loud, but not in English. At the same moment, I looked inside his cinched hood and recognized him.

It was Vlad. He grinned at me through the glass, then reached into his ragged pants and pulled out a gun.

Andrei popped up from the footwell and looked straight at Charlotte. "DRIVE! JUST GO!"

Frantic, she hit the gas. The Tahoe leapt forward, tires squealing.

The acceleration threw me back, and my head banged against the edge of the passenger seat. At the same moment, I heard a tremendous CLANK against the SUV's door, less than a foot away from me. Vlad was shooting at us.

"*Oh my God, oh my God!*" Charlotte stared straight ahead in terror. The Tahoe barreled toward the intersection where 82nd Street made a T-junction with Central Park West. She needed to turn either left or right, but she seemed paralyzed, her hands frozen on the steering wheel. We were just seconds away from smashing into the stone wall that surrounded Central Park.

I raised my head to get Charlotte's attention. "Make a right! *A right!*"

At the last second she jerked the wheel, and the Tahoe skidded into the southbound lanes of CPW. I'd prepared for the turn, bracing myself against the dashboard, but the centrifugal force tossed Andrei across the backseat. I heard another CLANK against the car's chassis—a second shot from Vlad, who stood in the middle of 82nd Street—but then we rounded the corner and careened out of his line of sight.

I dared a look out the Tahoe's rear window and saw a man in a pinstripe suit run past the main entrance of Dr. Cauchon's building. I couldn't tell if this agent carried a gun or not, but I wasn't about to take any chances.

"Make another right!" I yelled at Charlotte, pointing at the corner of Central Park West and 81st. "Right here!"

She made the turn, her fingers white against the steering wheel. Then we raced west on 81st Street, weaving around the taxis and buses. The Tahoe hurtled toward Columbus Avenue.

Andrei braced himself by clutching the driver and passenger seats from behind. He looked out the SUV's rear window, obviously scanning the street for more agents. After a few seconds he turned to me, his face sagging with relief. "All right, I don't see any others. I think we're safe."

Charlotte shook her head. Her cheeks were wet and her chin was quivering. "Who was that guy? And why was he shooting at us?" Her voice was loud, high-pitched, hysterical. "What's going on, Joan? What have you gotten us into?"

I reached for her shoulder and clasped it. I felt so awful at that moment, I almost started crying again. But I managed to hold it together and smile at Charlotte. With my other hand, I pointed at the street ahead. "Make a left on Columbus, okay?" I kept my voice low and squeezed her shoulder. "Then we're gonna make a right on 79th. We'll get on the West Side Highway and go across the George Washington Bridge."

Andrei leaned forward, poking his head between the seats. "And then where?"

The answer was the Science and Religion conference in New Mexico, but I was afraid to say it. Somehow I sensed what was going to happen to me there.

It wasn't a God-given prophecy or anything like that—I couldn't see the future or predict any specific events. I just had a bad feeling in the pit of my stomach.

So I did the cowardly thing. I dodged the question.

"We have a job to do. The most important job you can imagine." I nodded, confirming the justness of our cause, the rightness of our mission. "God wants us to save the world."

Part Two: Revelation

Chapter Nineteen

We drove west without stopping for the next four-and-a-half hours. After our narrow escape from New York, I wanted to get far away from the city, *fast*.

The first hour of the trip was the hardest. We were hyper and jangled and completely traumatized. Charlotte kept asking me questions, but she was too dazed to listen to my answers. Half a dozen times I tried to tell her the story, talking slowly and patiently while she drove the Tahoe down the highway, but she interrupted me every time after just a couple of sentences. "What? Wait, what are you saying?" She frowned and shook her head and strangled the steering wheel. There was so much fear and confusion on her face. It made me feel horrible.

It was all my fault. I should've never dragged her into this.

Andrei wasn't much better. As we drove from the George Washington Bridge to the New Jersey Turnpike, he spent most of the time staring out the rear window of the SUV. He insisted that Charlotte and I power off our cell phones, and he scrutinized all the cars traveling behind us. He became especially agitated when we got off the turnpike and switched to I-78; as we slowed down and approached the tollbooths, he swiveled his head back and forth, inspecting every vehicle in sight, clearly on the lookout for anyone who might be pursuing us. Although he was half-dead from fatigue, he couldn't stop searching.

The tension didn't subside until we'd driven sixty miles across New Jersey and reached the Delaware River. We all relaxed a bit as the Tahoe crossed the bridge and cruised into Pennsylvania, maybe because now there was a whole state between us and the bad guys. Andrei stopped looking for Russian agents, and Charlotte stopped interrupting me, and I finally got the chance to tell them about God.

All in all, Charlotte handled it pretty well. She'd already seen me preaching and crusading on YouTube, so she wasn't shocked when I said I'd received specific instructions from the Lord, commanding me to go to the conference in Albuquerque. The details of string theory confused her, but when I told her that God had revealed the true shape of the universe to me, she got the gist of it. She stayed calm even when I told her about Vlad. In fact, she seemed relieved to find out why someone had shot at us. The memory of it still terrified her—she winced as I described who Vlad was—but at least she had an explanation now. She muttered, "My God," and shook her head again.

Andrei listened to the whole story too, but he didn't say a word. When I was done talking, I looked over my shoulder and saw him sitting in the middle of the backseat, rigid and blank-faced. His eyes were open but he didn't seem to see me at first. I whispered, "Andrei?" and after a few seconds he nodded.

"Sorry. I need to close my eyes." Then he stretched out across the seat and went to sleep.

By this point we were on the Pennsylvania Turnpike. Charlotte asked me to find some decent music on the radio, and at first I had no luck. All I could get were country music and Top 40 stations, one after another. But eventually I found a college radio station that played the music I liked, the weirder stuff. First they did an indie-rock set—Arcade Fire, the Lumineers, Halsey. Then they slowed things down with an oldie by the Indigo Girls. And then, to my delight, they played "She Keeps Me Warm" by Mary Lambert.

That was my favorite song back when I was in middle school. I asked Charlotte if she knew it, and she said it sounded kind of familiar, and then I started gushing about how much I loved it. I didn't mention the fact that the song is about lesbians, but the lyrics made the subject pretty clear, and I thought maybe it would get us talking about emotions or relationships. Charlotte, though, didn't say very much. I realized she was probably overwhelmed and didn't want to think about *anything* for a while. So I shut my mouth, and we just listened to the music.

But I would've preferred conversation. As soon as we stopped talking, my mind wandered back to Manhattan, back to the moment when I burst out of Dr. Cauchon's office and ran past my parents without saying a word.

When did they realize that I'd slipped out of the psychiatrist's apartment? After five minutes? Ten minutes? And did Mom and Dad hear the gunshots when Vlad fired at our SUV? Did they call the police to report me missing? Or go to the precinct station to file a report? I could imagine it all so clearly, their desperate faces, their anxious questions, the terrible things they would say to each other.

And again, it was all my fault. I shouldn't have done it. Not even to save the world.

* * *

By 6 p.m. the Tahoe was running low on gas, so we stopped at one of the turnpike's service plazas. Andrei woke up from his nap and went looking for the restrooms, and I gave Charlotte sixty dollars to pay for the gas and a few

Arby's roast beef sandwiches too. Then I got out of the car and stepped toward the gas pumps.

That's when I noticed the bullet holes in the SUV—one jagged ellipse in the passenger-side door and another in the rear bumper. My pulse started racing, and not just because the holes reminded me of the shooting. The damage looked suspicious. If a state trooper got a good look at the Tahoe on the highway, he'd definitely pull us over, and it wouldn't take him long to figure out that Charlotte had commandeered her stepmother's rental car. Then our journey would come to a very abrupt end.

So I gassed up the SUV as quickly as possible, then leaned against the door to hide one of the bullet holes. Charlotte and Andrei soon returned with the sandwiches, chatting with each other as they approached the Tahoe, both of them looking more cheerful and less traumatized now. I stepped to the side and discreetly pointed out the damage to the car.

"This is a problem." I kept my voice low, even though no other motorists were nearby. "I think we should park the car in the farthest corner of the parking lot and stay there until the sun sets. The holes will be less noticeable in the dark."

Andrei thought it over for a moment. "We should also tape some cardboard over the holes. Then it would look like we were covering up a couple of dents, which would be more normal, yes?"

Charlotte smiled and clapped him on the shoulder. "Yeah, that's a good idea!" She pointed at the service-plaza building where she'd bought the Arby's sandwiches. "There's nothing but fast-food places here, but maybe we could slip a few dollars to one of the employees? And they could find a cardboard box and some tape for us?"

"Okay, that brings us to another problem." I reached into the pocket of my jeans and pulled out my wallet. "I have less than eighty dollars left, and this SUV is a gas guzzler. How much money do you two have?"

Andrei dug into his pockets and came up with a ten-dollar bill and two crumpled fives. Charlotte had a small yellow purse hanging from her shoulder, and although it contained only a handful of change, she smiled as she reached inside it. "Don't worry about the money. I know where we can get some more." She removed a folded slip of paper from her purse. "This is Marguerite's address. Remember Sister Marguerite? My online friend? The Everlight who wrote the righteous comment about your YouTube video?" She closed her purse and unfolded the slip of paper. "Before we left New York, I sent her a text. I told her I was going to drive across the country with you, and maybe I'd have a chance to finally meet her in person."

I remembered Marguerite. She was the YouTube watcher who'd informed the entire universe that the star of "Crazy God Girl in NYC Subway Station" and the flag-bearer in "Crusader on Second Avenue" were one and the same. I also remembered what Charlotte had said about her: *We talk online almost everyday, so we're pretty close.*

I pointed at the paper in Charlotte's hand. "Where does she live?"

"Alverton, Pennsylvania. There was a map of the Pennsylvania Turnpike on the wall next to the Arby's, and I found Alverton on it. It's near exit 91, about an hour from here."

"And you think we can just show up at your friend's house and she'll give us money?"

Charlotte nodded. Her face was the very picture of confidence. "Marguerite is a true believer. When my dad created the website for the Church of Everlasting Light, he set up a page for making donations, and I've been running it for the past year. So I know that Marguerite donates two hundred dollars to the church every month."

Andrei nodded too, evidently impressed. "Do you think your friend might let us stay at her house overnight? I'd love to sleep on a bed tonight instead of this car's backseat."

"Of course!" Charlotte clapped him again, and this time her hand lingered on his shoulder. "The Everlights are more than just a church. We're an online community of faith. Even though we live thousands of miles apart, we're always ready to help each other. That's our most important rule."

I stared at Charlotte's fingers as they pressed against Andrei's polo shirt. I'd worried about whether they'd get along with each other during our cross-country drive, but now it looked like they were getting along a little *too* well. It occurred to me that maybe I'd made the wrong assumption about Charlotte's sexual orientation. Ever since she touched me inside the Church of Saint Paul, I'd assumed she was lesbian, but now I started to wonder. Maybe she was bi. Or maybe even straight.

No! She's attracted to me! I felt it so strongly!

I shook my head, trying to dispel the upsetting thoughts from my brain. Then, turning away from Charlotte and Andrei, I opened the Tahoe's door. "All right, we can't just stand here. Let's move the car to the parking lot."

My anxiety must've shown on my face, because Charlotte suddenly let go of Andrei and stepped toward me. She clasped my right arm above the elbow—ever so gently—and as soon as she touched me, I felt the attraction again, that warm promise. If Andrei hadn't been there, she probably would've kissed me. Instead, she leaned very close and looked into my eyes. "And after it gets dark, we'll go to Marguerite's house? You trust me, right?"

I nodded. And smiled at her. I was in love. There, I said it.

* * *

A waning moon rose above the Pennsylvania hills as we approached the town of Alverton. The Tahoe's headlights shone on Route 981, a deserted country road that snaked between cornfields and dairy farms. Charlotte squinted at the road ahead, trying to read a mileage sign at the next intersection, her face silvered by the dashboard's glow.

"Hey, look! We're only two miles away!" She pointed at the sign. Her voice was ecstatic. "I knew we were getting close!"

I wasn't nearly as happy about it, but there was no point in making a fuss. We needed help to get to New Mexico, and we couldn't be too picky about our benefactors. "Before we get there, can you tell me a little more about Marguerite? Do you know how old she is?"

"She's a year older than us, I think." Charlotte slowed the SUV and tapped her index finger against the steering wheel. "Yeah, she's probably eighteen, because she texted me a picture of her graduation last June. She had a really rough time in high school. Both her parents died during her junior year."

Andrei leaned forward from the backseat. "How did they die?"

"A car accident. Afterwards, Marguerite and her older brother got some life-insurance money, a whole lot of it actually, more than enough for them to live on. But her brother is a real creep, a guy in his twenties who never leaves the house and spends all his time playing video games in his bedroom. She can't stand him."

I was starting to understand why this girl and Charlotte had become online friends. Both of them had lost their parents. Both had been forced to live with someone they hated. "So is that why she joined your church? Because she was unhappy?"

Charlotte shook her head. "No, not at all. Marguerite is an idealist. She turned her tragedy into something wonderful. Because she has all that insurance money, she can help out so many people in need of healing. And ever since she graduated, she's devoted herself full-time to her spiritual causes. She's been an Everlight for a long time, but she's also a big supporter of the Cyber Shaman Temple."

"Cyber Shaman?"

"You never heard of it? It's one of the most popular websites for religious seekers. Their doctrine combines Tibetan Buddhism with Native American traditions. Marguerite is an important leader in that online community too."

I said nothing in response. I didn't know what to think of the whole phenomenon of Internet spirituality. I certainly wasn't prejudiced against the smaller religions, the ones that didn't have massive cathedrals or monasteries, but there were so many creeds and cults on the Web, too many to keep track of. So I reserved judgment about the Cyber Shaman Temple and Marguerite. Maybe she would turn out to be perfectly sensible.

Soon we cruised into Alverton, which was nothing more than a few dozen homes and trailers lined up alongside Route 981. At the far end of the town, Charlotte steered onto a narrower road that climbed a wooded hill, and then she started looking at the mailboxes and trying to read the address numbers. In preparation, I reached under my seat and picked up the math notebooks. After another half-mile Charlotte whooped, "Here it is!" and turned down a gravel driveway that curved behind the trees.

At the end of the driveway was an ordinary two-story house. It had a big front lawn and a couple of illuminated windows on the second floor and a picture window with closed curtains on the ground floor. Even in the darkness, though, I could see signs that something was wrong. A minivan was parked in the driveway but it was in terrible shape, with large patches of rust above the wheels and a banged-up door gleaming in the moonlight. The lawn was packed with knee-high weeds and conical mounds of metallic junk, and the house's roof seemed to be missing a few shingles. In short, it didn't look like the kind of place where rich people lived. I started to doubt whether anyone here could help us.

Charlotte parked the Tahoe behind the minivan, shut off the engine, and stepped out of the car. As she headed toward the house, I hopped out of the SUV with the notebooks under my arm and intercepted her. "Listen, can I ask you a favor? Let me do the talking. I want to get to know Marguerite before I tell her what we need and where we want to go. We have to be careful what we tell people, you know?"

I thought Charlotte might get offended on her friend's behalf, but she was all smiles. "Sure, no problem. You're the best person to talk about it, anyway." She hooked her arm around mine, and we walked side by side to the front door. "And Marguerite will be totally stoked to meet you. She's been spreading the word about your videos, you know. She forwarded the links to all her chat groups."

Andrei had exited the Tahoe by this point, and he followed us to the house. When we got to the porch, Charlotte raced up the steps and reached for the doorbell, but just as she was about to ring it, the door opened wide. A tall girl stood in the entryway, dressed in gray sweatpants and a bulky black sweater. She looked nervously at the three of us.

Charlotte stepped forward. "Marguerite? Hey, it's me, Charlotte. I—"

Before she could say another word, Marguerite rushed outside to hug her. The impact almost knocked Charlotte over. I felt it in my own chest, even though I stood several feet away.

"Charlotte! I can't believe it! You're really here!" Marguerite's voice was loud and husky. It boomed across the front lawn. "This is so glorious!"

She was a big girl, at least six feet tall. She towered over Charlotte as they hugged, her chin jutting over Charlotte's head. Although Marguerite was white, she'd twisted her hair into a dozen brown dreadlocks that stretched halfway down her back. She had a pale, round face and a joyous smile, and as she wrapped her arms around Charlotte I felt absolutely awful. I had no idea if Marguerite was lesbian, but the possibility petrified me.

Charlotte said something, but I couldn't hear it. Her face was mashed against the crook of Marguerite's neck, so her voice was muffled. After a second, though, she tilted her head upward. "Didn't you get the text I sent this morning? I told you I might be coming."

Marguerite nodded. "I got it, but I didn't expect you so soon. Were you driving all day?"

"Yeah. We had to leave New York in a real hurry." Charlotte stepped backward, slipping out of Marguerite's embrace, and pointed at me and Andrei. "This is Joan, the girl I told you so much about. Joan's in charge of our little expedition, and she's going to tell you all about it. And that's Andrei, her friend."

Marguerite ignored Andrei and stepped toward me. She grasped my right hand with both of hers and leaned over, bending her knees to bring her head down to my level. It was such a deferential posture that it weirded me out a little. It felt as if she were kneeling at my feet.

"Joan, this is an honor. Charlotte told me about your revelation on the subway, and I must've watched the video a dozen times. And when I saw the second video this morning, I knew beyond a doubt that a miracle was happening." She squeezed my hand between hers, clamping it tight. "I'm so proud to be a part of your blessed mission, whatever it is. Please, come inside my home and let me get you something to eat."

Keeping her grip on my hand, she escorted me through the doorway. Charlotte and Andrei followed us, a few steps behind.

Marguerite led me into the messiest living room I'd ever seen. A long beige sofa ran along one of the walls, but it was loaded with so many piles of newspapers that there was no place to sit. Flanking the sofa were two reclining chairs occupied by stacks of junk mail, and the coffee table was crowded with greasy plates and bowls. On the other side of the room was a television set with a cracked screen, and on the floor was an off-white carpet spotted with brown stains. A couple of flies buzzed over the sofa, and a few more crawled across the dirty dishes on the coffee table.

The place was a disaster area, and yet Marguerite didn't seem the least bit embarrassed. She tossed the stacks of newspapers off the couch and swept the junk mail off the recliners. "Okay, everybody, sit wherever you like. Now, how hungry are you? I've already had my dinner, but I could heat up some tomato soup."

I took a seat in one of the recliners and rested the math notebooks in my lap. Andrei and Charlotte sat on the far end of the couch, trying to keep their distance from the sticky mess on the coffee table. I started to wonder again if Marguerite could really help us—I found it hard to believe that a philanthropist lived here—but I pushed that thought aside. The important thing was that Marguerite and Charlotte were friends. Whether she could help us or not, I was going to treat her with respect. "That's such a nice offer, but we bought some sandwiches an hour ago and ate them in the car."

"Oh, that's too bad! Are you sure you don't want anything?" Her voice rose in pitch. She was so eager to please, it was making everyone nervous. "Maybe some cookies or crackers? Or some ice cream?"

Charlotte and Andrei shook their heads, and I did the same. "No, we're fine. But thank you so much for inviting us into your home."

"I'd offer you some tea, but we ran out of teabags. I asked Ben this morning to pick up some more at the supermarket, but he forgot. Ben's my older brother, he's a video-game addict." Marguerite pointed a finger straight up. "He's upstairs in his bedroom now, and he probably won't come down. All he ever does is play his shoot-'em-up games, so he's not very sociable." She sat down in the other recliner. "But I've always been a big believer in hospitality. Like the Bible says, 'Do not neglect to show hospitality to strangers, for by doing this some have entertained angels without knowing it.' That's a lovely verse, don't you think?"

Marguerite was talking a mile a minute. I couldn't tell if this was natural for her, or if it was a side effect of her nervousness. I smiled, trying to put her at ease. "Yes, it's beautiful. You seem to know the Bible pretty well."

"I grew up with it. My mom and dad were strict Baptists—church every Sunday, Bible study every Wednesday night. But Charlotte told you what happened to my parents, right?" She glanced at Charlotte, who nodded very seriously. Then Marguerite turned back to me. "After they died, I realized their religion was too small for me. It couldn't answer all my questions or hold all my heartbreak. So I did some research on the Internet and discovered a larger world of spirituality. I had a vision of a new kind of God, one who was utterly transcendent, so vast and deep that no book could describe it. And I found hundreds of other people who shared that vision." She rose halfway out of her recliner and pointed at me. "It's the same God you saw in *your* vision, Joan. The God of energy, the God of transformation, *the God of fire!*"

Marguerite's face reddened, especially her forehead. Her mouth hung open and her dreadlocks swayed and her outstretched arm trembled above the coffee table. It was a little scary to witness her passion. It rose inside her so quickly, animating every part of her body. Under ordinary circumstances, I would've worried about her mental state and maybe tried to calm her down.

But instead I nodded. The passion was inside me too.

"That's right, Marguerite. God appeared to me three times, first in the form of a boy, then as a homeless woman, then a police officer. And She showed me something very special." I opened the math notebook on my lap and turned to the sketch of the Calabi-Yau manifold. The page was starting to look ragged because I was showing it off so much. "You see all the loops and twists and folds? If you could somehow observe the tiniest volume of space, more than a trillion times smaller than a proton or a neutron, this is what it would look like. And this shape determines the properties of everything in the universe—matter, radiation, electricity, gravity. Basically, you're looking at the cosmic blueprints. It's the hidden geometry behind everything we see."

Marguerite stood up and stepped toward me, walking slowly across the living room. Her eyes widened as she stared at the manifold. "That was part of your vision? What the Lord revealed to you?"

I nodded again. "It's called the Theory of Everything. God wants everyone to see it, because it's Her grand design for the universe. She told me to reveal the theory to the scientists and the peacemakers, because it's the only thing that can prevent a terrible war from happening. The revelation will stop the killing and bring lasting peace to the world."

Marguerite bent over to take a closer look at the sketch. Then she did what I was afraid she'd do: she clasped her hands together and knelt on the carpet. Her big brown eyes swelled with tears.

"Oh precious God! Oh most righteous and loving Preserver of Creation!" She looked up at the ceiling, then closed her eyes. Tears streamed down both sides of her face. "Thank you, Lord, for this blessed gift! And thank you for bringing your holy messenger to my home!"

I had to look away. I was ashamed of myself. Even though I hadn't lied to Marguerite—I'd told her the gospel truth about my visions—I still felt like I was using her. I was exploiting her faith, her simple guilelessness.

But how else could I get to New Mexico? How could I follow God's commands and save the world from catastrophe if I didn't embrace the believers She placed in my path?

Uncomfortable, I averted my eyes from the kneeling girl and turned toward the sofa. Charlotte was crying too, in concert with her friend. She wept silently and brushed the tears from her cheeks. But Andrei just sat there, frowning. His face was a haggard mask of disapproval.

I turned back to Marguerite. "Okay, here's the thing. I need to get to Albuquerque, New Mexico, by Sunday morning. I'm supposed to reveal the Theory of Everything at an important Science and Religion conference that's taking place this weekend. I wasn't officially invited to the conference, so I don't know how I'm gonna get a chance to speak there, but I can't worry about that yet. Right now I'm just worried about getting there, because we don't have enough gas money for the trip."

There it was. I'd made my pitch. I felt like one of those TV evangelists who appear on the cable channels late at night, the con artists who preach about redemption and healing and then urge their listeners to dig deep into their pockets and call the telephone number at the bottom of the screen. I expected Marguerite to run upstairs and come back down with a jar of pennies for me.

Still kneeling on the carpet, she opened her eyes. "Joan? I've heard about this conference. It's called 'Bridging the Gap Between Science and Religion,' right?"

This was a surprise. I closed my math notebook. "Uh, yeah, that's right. How did you hear about it?"

"Oh, it's a big deal. Everyone's talking about it on the Web." Marguerite got off her knees and stood up in front of me. "That conference is the biggest topic right now on all the spirituality forums. They're calling it GapCon. Here, let me show you." She grabbed a laptop that lay on the coffee table next to the dirty dishes. "A lot of folks are upset because they think GapCon will be biased against religion. Most of the scientists invited to the conference are atheists, people like Richard Dawkins and Daniel Dennett. You've heard of them, right? They've written books saying that God is a fantasy and that all religious believers are deluded idiots."

Marguerite opened the laptop and typed a Web address on the keyboard. Then she crouched next to my recliner and held the computer in front of me. On the screen was a website called "Mystery and Metaphysics." The headline on the site's home page was, "Heavenly Fury: Spiritual Leaders Around the World Denounce GapCon."

I leaned closer to the screen, but the text below the headline was tiny. "Spiritual leaders? Who are they talking about, the pope?"

"No, he won't get involved in this kind of battle. The Catholic Church is focusing its energy elsewhere. It's a real shame, because Catholic ritual and theology are so inspiring." She shook her head in disappointment, her dreadlocks swinging. "And forget about the Protestant churches. Most of them are too scared to fight. The only congregations brave enough to challenge the atheists are the online communities. You know, like the Wiccan Web and Cyber Shaman and Everlasting Light."

Charlotte got up from the sofa and came over to my recliner. She stood behind us and looked over Marguerite's shoulder at the laptop. "Does that article mention the Everlights?"

Marguerite shook her head again. "No, but do you know Tiffany Morningstar at SaviorNet? The article says she's organizing a protest against GapCon. They're gonna gather outside the convention center in Albuquerque where the conference is going on."

I felt a gentle pressure on the back of my arm, just below the sleeve of my T-shirt. Charlotte had touched me to get my attention. "Joan, don't you see what's happening? All the pieces are falling into place. By the time we get to New Mexico, the whole world will be paying attention to the conference. Everything will be ready for your revelation. It's all part of God's plan."

Marguerite nodded vigorously. She stood up, closed her laptop, and put it back on the coffee table. Then she stretched her right arm toward Charlotte and her left toward me. We formed an isosceles triangle, with me at the vertex, sitting in the recliner as if it were my throne. As if I were a queen and the two Everlights were my loyal subjects.

Marguerite squeezed my shoulder. "Yes, God wills it! We'll travel together to GapCon!" She nodded a few more times, then smiled at me. "And you know what? I'll get on my network right now and see if anyone else can join us! We'll turn this into a pilgrimage!"

I leaned back in the chair, pulling away from her. Maybe this was part of God's plan, but it wasn't part of mine. "You want to come with us? To Albuquerque?"

"Why not? I still have three thousand dollars in the bank, and that'll be enough to pay for the gas and everything else." Marguerite let go of my shoulder but kept smiling at me. "And maybe we should take my minivan instead of your SUV. My car probably gets better gas mileage."

Charlotte edged closer, leaning over the arm of the recliner. She clearly sensed my reluctance. "This is a good thing, Joan, a really good idea. I mean, think about it. A minute ago you said you were worried that they wouldn't let you speak at the conference. But what if you came to Albuquerque with a big group of supporters backing you up? Then those scientists and atheists couldn't just ignore you. They'd have to give you a chance to speak, right?"

There was a flaw in her logic, of course. If my supporters were a bunch of loony shamans and zealots and New Agers, they'd hurt my cause much more than they'd help it. But I couldn't make this point while sitting in Marguerite's living room. It would crush her feelings and hurt Charlotte too.

So I took a different tack: I turned away from the Everlights and focused on Andrei instead. He still sat at the end of the sofa, looking glum and disgusted.

"Andrei, what do you think? Are you okay with Marguerite coming with us?"

I expected him to say no. I could tell from his body language that he wasn't comfortable in this company. He might put up with Charlotte for a while, because she was pretty and sweet and relatively normal, but Marguerite was a different story. Like me, Andrei was a mathematician, and mathematicians hated disorder.

But to my great surprise, Andrei nodded. "The fact is, we have to travel in her minivan anyway. We can't go any farther in the SUV."

"Why? We covered up the, uh, damage. The car looks normal now, more or less."

Andrei looked at Charlotte. "It's a rental car, isn't it?"

"Well, my stepmother rented it, but I have all the paperwork from the Avis rent-a-car place, and I left her a note—"

"Is there a chance that your stepmother will report it missing? To the police?"

Charlotte thought it over. "Yeah, actually she might. Especially when she figures out I'm not coming back to her hotel tonight. That's definitely gonna tick her off."

Andrei turned back to me. "It's an expensive vehicle, so it's equipped with a GPS tracking system. And Avis will activate the system and locate the SUV as soon as they get the police report."

"So we need to ditch the Tahoe before that happens?"

He nodded. He said nothing else, but his message was clear: we didn't have a choice. We had to team up with Marguerite, whether we liked it or not.

Which was disturbing. I had the feeling that I was being manipulated. God was arranging things for us without our knowledge or consent. And that made me uneasy.

Nevertheless, I put a smile on my face. We had a mission to accomplish, and nothing else mattered. I rose from the recliner and looked Marguerite in the eye, ready to tell her how grateful I was for her offer to drive us to New Mexico.

But then a tall man in a red tracksuit stormed into the living room. He rushed past the television and the coffee table, heading straight for us, his sneakers pounding the stained carpet. His face was sweaty and scowling, and in his right hand he held something shiny and black.

I froze. I couldn't move.

The Russians. They found us.

But the man was much too young to be Vlad—he couldn't have been more than 24 or 25—and as he stretched his right arm toward us, I saw that the object in his hand wasn't a gun. It was a video-game controller, and he was pointing it at Marguerite.

"All right, I've heard enough! You can't pull this crap, Margo!"

His voice was nasty. He sneered at us, curling his upper lip, which was flecked with saliva. His cheeks and chin were dotted with stubble, and his hair was black and matted. His tracksuit was hopelessly wrinkled, as if it had been lying on the bottom of a closet for months, and his sneakers were full of holes. Clearly, he didn't get out much. Marguerite stepped forward to confront him, and at that moment I guessed who he was: her anti-social, video-game-addicted brother.

"Ben? You were listening?" Marguerite's face reddened again, and she balled her hands into fists. "*You were spying on us?*"

"Yeah, and I'm glad I did. You think you can blow all our money on some *pilgrimage*?" Ben swung his arm from left to right, pointing his video-game controller at Charlotte and Andrei and me. "Who are these people, anyway? You don't even know who they are, and you're gonna give 'em three thousand dollars?"

Marguerite didn't back down. She stood a yard away from her brother and glared at him. "First of all, it's *my* money! We split the insurance settlement *in half*, remember?" She spoke with exaggerated slowness, as if talking to a child. "You can spend *your* half any way you like, and I can do whatever I want with *mine*!"

"We also have to split the expenses, idiot. The electric, the gas, and the freakin' cable bill." Now he swept the controller in a circle above his head, gesturing at the whole room. "You've already thrown away a fortune, sending donations to every nutcase religion on the Internet. So what are you gonna do when you're broke? How are you gonna pay your share of the bills?"

"I'll get a job! Unlike you, I'm not afraid of working!"

Ben chuckled. "Oh yeah? Who in their right mind would hire you? You're the screwiest girl in town, and everyone knows it."

Marguerite spun away from him as if she'd been slapped. She glanced at Charlotte and me, then raised her hands and clamped them to her head, one on each side. She dug her fingers into her dreadlocks and let out a scream, a piercing cry of humiliation. Then she raced out of the living room and into the kitchen.

Charlotte ran after her. I felt a powerful urge to follow them and an equally powerful urge to knock the stuffing out of Marguerite's brother. But before I could act on either impulse, Andrei jumped off the couch and approached me, coming close enough to whisper in my ear.

"Let's go outside. We need to talk."

Meanwhile, Ben ignored us. He took a seat in one of the recliners, picked up a remote control from the filthy coffee table, and turned on the television. I decided to ignore him too and go with Andrei, making sure to take the math notebooks with me.

Just as I was leaving the house, though, I looked over my shoulder. Ben waved goodbye at us.

"It was nice meeting you, wackos. Don't let the door hit you on the way out."

Chapter Twenty

It wasn't until we stepped outside that I noticed how scared Andrei was. He stood on the front lawn, breathing fast and looking up at the night sky. His face was almost as white as the moon.

Alarmed, I grabbed one of his hands. It was very cold. "What is it? What's wrong?"

His teeth chattered. "That man. The brother." He tilted his head toward the house. "When he came into the room like that, so quickly…"

Andrei's voice trailed off, but I understood. "You thought he was a Russian agent? One of Vlad's men?" I squeezed his hand. "I thought the same thing."

He nodded. "What he wore? That tracksuit? Many Russians like to wear those. Especially the Russian gangsters. They think it makes them look powerful."

"Really? That's weird. It looks ridiculous."

Andrei kept nodding. Ordinarily, I would've worried that he might get romantic thoughts in this kind of situation—I was holding his hand, after all, and the moon was shining—but his mind was clearly elsewhere. He was freaked out.

After a few seconds, he pulled his hand out of mine and marched over to where the Tahoe was parked. Bending over, he peered through the driver-side window. "Good. Charlotte left the keys in the ignition." He opened the door. "Would you like to come with me?"

"Whoa, wait a second. What are you—"

"I'm just going to move the vehicle. We passed a parking lot on the road, not so far away, maybe a kilometer from here. Ditching the car in that lot would be better than leaving it in front of this house. That way, when the police locate the SUV, they won't make the connection between us and your friend Marguerite."

I thought it over, and it made sense. So I got into the Tahoe's passenger seat, and Andrei got behind the wheel. He managed to smile as he started the engine. "I have to warn you, I don't have an American driver's license. But back in St. Petersburg, my father owned a Renault." He shifted the SUV into reverse. "This car is a little bigger, but the principle is the same."

Andrei proved his skill by expertly backing out of the driveway. He got on the narrow country road and drove down the hill, retracing the route we'd taken to Marguerite's house. He was breathing normally now, but I could tell he was still agitated. He hadn't fully explained what was worrying him.

"Talk to me, Andrei. What's bothering you?"

He said nothing for several seconds. Then he swallowed hard. "Sooner or later, Vlad will find us. He's smarter than the police."

"What do you—"

"His agency has spies all over this country. And he has the software that my father designed, the code-breaking program. He can use it to hack into any government network, including the computers used by local and state police departments."

"Seriously? He can access their databases?"

"He can monitor the video from every one of their surveillance cameras. And he can break into the networks of private companies too. That's why I told you and Charlotte to power off your cell phones. If you turn them on, AT&T will know where you are, and so will Vlad."

I shook my head. "But wouldn't it be a huge risk for him, hacking so many computers? I mean, are we really worth all that effort? Doesn't Vlad have better things to do than chase us?"

Andrei paused again before answering. He made a right turn, putting the Tahoe back on Route 981. "I can tell you one thing for certain, Joan. Vlad's agency once used a radioactive isotope to poison one of its enemies. They don't give up easily." He grimaced. "My father didn't realize how dangerous they were. Not until the end."

He drove the rest of the way in silence. After another half-mile, Andrei turned into the parking lot he'd seen earlier and steered the Tahoe into a space in the lot's far corner. I took the math notebooks with me as I got out of the car, but Andrei left the keys in the ignition. It was a clever move—if someone stole the Tahoe, it would lead the police farther away from us.

Then we headed back to Marguerite's house, walking side-by-side on the shoulder of Route 981.

It was ten o'clock on a Thursday night, and the town of Alverton looked utterly deserted. We made quick progress at first, passing a darkened school, a shuttered gas station, and a fenced-off junkyard. But our pace slowed after we reached the turnoff to the narrower road, because now we had to climb uphill.

Andrei walked with his head down and his hands in his pockets. He'd clearly lost interest in talking, so I just stared at him for a while, sneaking looks out of the corner of my eye. He seemed so unhappy compared with the boy I'd met three days ago, the goofy, grinning kid who'd wanted to study string theory with me. And wasn't I responsible for his unhappiness? If I hadn't called him up the night before, if I'd just kept the Theory of Everything to myself, then Andrei wouldn't be in so much trouble now. He would still be in New York, at his new high school, still grinning and acting goofy and trying to forget what had happened to his father. But I'd ruined his new life and brought all his bad memories back. I'd made it impossible for him to forget.

So I decided to apologize. I edged closer to him as we walked along the dark road.

"Listen, I'm sorry for the way things turned out. I didn't know everything would get so…so…" I stopped, unable to think of the next word. I needed to start over. "I mean, I know this trip to New Mexico must seem ridiculous to you, but I—"

"No, it's not ridiculous. Not at all." He shook his head firmly. "The conference in Albuquerque will be a good place to present your work. I assume there'll be many physicists and mathematicians there?"

"Oh yeah, hundreds of them."

"That's excellent. They'll analyze your solution and look for flaws and give you immediate feedback. And because you'll explain the theory to hundreds of scientists at the same time, no one will be able to steal the idea from you and take credit for the discovery."

I nodded, pleased. "Yeah, you're right. I hadn't even thought about that. Although in all fairness, I don't deserve credit for the solution either. God showed me the most important part."

Andrei suddenly turned his head sideways and looked at me. "Okay, that's where we have a difference of opinion. I don't see the need to insert God into this discussion."

I was so surprised, I laughed. "Come on, we went over this. I explained it to you and Charlotte in the car, remember? The homeless woman on the subway? How she stopped time and everything froze, and there was a pattern of frozen light around the moon?"

"Let me propose a hypothesis. You were tired that night, so you fell asleep in the subway car. You dreamed about the homeless woman and how she made time stand still. And as part of this dream, your brain took a creative leap and drew the Calabi-Yau manifold. Isn't that a more likely explanation?"

"No, it's *not* likely. It's absurdly improbable." I raised my voice, because I was frustrated. I'd discussed probabilities at great length while we were driving on the Pennsylvania Turnpike, and I'd assumed that Andrei had been paying attention. "Remember the number of possible Calabi-Yau manifolds? Ten to the power of two-hundred-and-seventy-two thousand? How could I have guessed the correct shape out of all those possibilities?"

"Your guess wasn't random. You knew what you were looking for, and that improved your odds of success." Andrei pointed at my forehead. "What was your state of mind right before you got on the subway? You were thinking hard about the problem, correct? Considering all the possible solutions? You couldn't solve the puzzle with your conscious mind, but when you went to sleep your subconscious took over. And that part of your brain made the right connections and enabled you to see the answer."

"So you think I solved it in my sleep? The most difficult problem in the history of physics?"

Andrei took a deep breath. It sounded very loud on the empty road. "Are you familiar with August Kekulé? The nineteenth-century German chemist?"

"What does he have to do with—"

"He made a discovery while he was dreaming. He was trying to figure out the molecular structure of benzene, how the carbon atoms linked together. He dozed off one evening and saw chains of atoms twisting like snakes. Then he saw one of the snakes curl around in a circle and bite its own tail. When he woke up, he realized he'd seen the correct structure of benzene: six carbon atoms arranged in a ring."

I frowned. It was a good argument, very scientific. Andrei had offered an alternative hypothesis to explain the evidence. And if the only evidence I had was my vision of the homeless woman, I might've given his hypothesis some serious consideration. But there was more.

"I wasn't asleep last night on Second Avenue. I was wide awake when that police officer put me in the back of his patrol car. And when He told me who He really was."

"All right, let's talk about that. This happened after you left Samovar, correct? After Vlad slapped you?"

I nodded but said nothing. The incident had humiliated me so much, it hurt just to hear Andrei mention it. For an instant I felt the pain and surprise again, the horrible shock.

"Joan, I'm so sorry. I should've gone with you when you ran after Vlad. But I was terrified of him, absolutely paralyzed. So I just sat there at the table." He turned away from me and looked at the other side of the road. "After five minutes I finally found the courage to leave the diner. I looked up and down the street, but you were gone. And I was worried, because Vlad had hit you in the head, and you seemed so shaky."

"Well, yeah, I got dizzy for a second, but it didn't last long."

"It was a vicious blow, very hard. And afterwards you weren't thinking normally. The fact that you chased Vlad out of Samovar is proof of that. And maybe there were other effects too. Maybe it disrupted your perception."

I grabbed Andrei's arm, stopping him in his tracks. At the same time, I held up the math notebooks. "Okay, here's evidence of my sanity. I got our notebooks back, didn't I? And there's a video on YouTube proving that a cop pushed me into his car. Is that convincing enough?"

Andrei didn't want to look at me, but I forced him to. He tilted his head to the side and bit his lip. It was a look of pity. "And then the police officer drove you home. You didn't seem dangerous to him, and it was easier to take you home than to arrest you. And maybe that's the whole story. No God, no revelation. Just a friendly officer who felt sorry for you."

I was furious. I let go of Andrei's arm and stepped away from him. "The cop told me about the conference in Albuquerque! Why would an ordinary New York police officer know about something like that?"

"Are you sure you didn't hear about the conference somewhere else? Perhaps you saw something about it on the Web?"

I took another step backward. I was so angry, I waved the notebooks in the air, their pages flapping. "Great! That's just great! You think I'm crazy!"

He held out his hands, palms up, in a pleading gesture. "No, Joan, I think you're a genius. You're the best mathematician I've ever met, and that includes all the teachers at my old school in St. Petersburg."

"So why do you keep fighting me? Why don't you believe what I'm saying?"

"Please, just listen. If you look at the history of mathematics, you'll see that the lives of many geniuses were difficult." He pointed at the night sky, as if the history of math was written in the constellations. "I don't want to frighten you, but you need to acknowledge the problem. Georg Cantor, the creator of set theory, died in a mental asylum. Kurt Gödel, the greatest logician of all time, thought someone was trying to poison him, so he starved himself to death. John Nash—"

I didn't let him finish. I screamed. The sound shredded my throat and shattered the darkness.

Then I started running. With the notebooks under my arm, I sprinted back to Marguerite's house.

* * *

I ran half a mile at full speed, uphill all the way. I didn't time myself, but I bet I did it in less than two-and-a-half minutes. I ran fast enough to burn off my anger, to boil all the fury out of my blood. Then I paced across Marguerite's weedy lawn for another minute, catching my breath and coming back to normal.

When I rang the doorbell, it was Charlotte who let me inside. She stood in the doorway, smiling, with a package of Oreo cookies in her hand. Given the tense situation in that house, she looked surprisingly cheery.

"Hey, I was wondering where you and Andrei went." She looked past me, surveying the driveway. "You moved the Tahoe somewhere else?"

"Yeah, it'll be safer that way." I pointed over my shoulder, acting casual, pretending that nothing was wrong. "Andrei should be coming back any minute."

"All right, we'll wait for him inside. Want an Oreo?"

She placed a cookie in my hand as she ushered me into the foyer. I smiled and thought of the first time we met, in the soup kitchen at the Church of Saint Paul the Apostle. She'd handed me a cookie then too.

I popped the Oreo into my mouth, then peered into the living room. To my relief, it was empty.

"Where are Marguerite and her brother?"

Charlotte rolled her eyes. "They started fighting again right after you left. Marguerite said she was going to New Mexico no matter what. Then Ben said if he couldn't stop her from going, he was gonna come with us."

"What?"

"Yeah, he's worried she'll do something stupid. I guess Marguerite has a history of falling for scammers. Ben is afraid she'll meet some Internet weirdo in Albuquerque and sign away their minivan or whatever."

"So he's going to be in the car with us the whole way down there?"

"Yeah, it sucks." She shook her head and stepped toward the living-room sofa. The television in the room was still on, but the volume was at the lowest setting, barely audible. "Marguerite got so upset about it, she locked herself in the bathroom. I stood in the hallway for a while, talking to her through the bathroom door, and after a few minutes she stopped crying. I told her she should relax, maybe take a bath, so that's what she's doing now. Her brother went back to his bedroom, probably to play some more video games."

I followed Charlotte to the sofa, and we both sat down on the middle cushion. "Wow, so much drama. You think God is testing us?" I smiled again. "Maybe She's throwing all this weird stuff at us just to see how faithful we are?"

This was a joke, sort of, but Charlotte took it seriously. She thought it over for a second. "Yeah, that would explain a lot. I'm starting to understand why everyone in the Bible was so miserable all the time." She reached into her package of Oreos and pulled out two more. After passing one of the cookies to me, she pointed at the television across the room. It was tuned to CNN. "We need a break. Want to see if something good is on TV? Maybe one of those old-fashioned tearjerker movies?"

My heart thumped. I knew I was probably being way too hopeful, but there was something in Charlotte's voice that sounded warm and inviting. And yes, maybe even a little flirtatious. We sat very close, our thighs almost touching. We were alone on the living-room sofa, at least until Andrei returned to the house, and she'd just said she wanted to watch a romantic movie with me. She leaned back against the cushions, grinning slightly, then stretched one of her bare feet toward the coffee table in front of us. "Can you get the remote? It's on the table, but I don't want to look at those disgusting plates again."

I grinned too. "No problem." I kept my eyes on her as I bent over the table and picked up the remote control.

Then I turned toward the television and saw an army on the screen.

I stared at the CNN news footage, a shaky video clip that was only five seconds long but was broadcast over and over again, at least six times in a row. It showed hundreds of soldiers in camouflage-pattern uniforms marching across a green meadow. Half a dozen gray tanks sped past them, their caterpillar treads tearing up the field, their big guns angled upward. At the bottom of the screen, in blinding red letters, were the words "RUSSIAN TROOPS INVADE THE BALTIC REPUBLICS."

My stomach twisted. I fumbled for the button on the remote control and turned up the volume. An unseen anchorman was providing a voiceover for the repeating footage.

"…so far there has been almost no resistance to the Russian army, which massively outnumbers the opposing forces in Lithuania, Latvia, and Estonia. A spokesman for Russia's Foreign Ministry denied that any invasion was taking place, but American forces in Eastern Europe have been placed on the highest alert. All three of the Baltic republics are members of NATO, which means that the United States is committed to defending them."

Charlotte leaned forward, squinting at the screen. "Oh Jesus. Oh my God. It's happening, Joan. Just like you said."

I dropped the remote. The anchorman kept talking about tank divisions and aircraft carriers and ballistic missiles and nuclear options, but after a few seconds I couldn't hear him anymore. He was drowned out by a remembered voice in my head, the words of God's warning.

Five billion people will be dead in two weeks.

Chapter Twenty-One

The minivan was a ten-year-old Honda Odyssey that had seen better days. Its chassis rattled and squeaked as Marguerite drove it across Ohio on Interstate 70. The engine was in even worse shape—it started to overheat if she drove too fast—and we had to stop every couple of hours to pour more coolant into the radiator. It was already one o'clock on Friday afternoon and we were still fifty miles from the Indiana state line.

Charlotte sat in the front passenger seat, tapping out texts and emails on Marguerite's iPhone, sending messages to everyone in their online communities. Between the two of them, they knew dozens of spiritual seekers on the Web, all pursuing their chosen faiths with electronic fervor. The Internet was particularly hysterical that afternoon as everyone reacted to the news from Eastern Europe. People were sharing videos, of course—the Russian army marching into Vilnius, a shouting match between the ambassadors at the United Nations—but Twitter and Facebook were also full of prayers for a ceasefire. So Charlotte, with her fingers dancing on the iPhone's screen, added our new gospel to the online clamor, spreading the story of our pilgrimage, our mission of peace.

Andrei and I sat in the minivan's second row of seats, our heads bent over the math notebooks in our laps. Now that we were on our way to New Mexico, our top priority was to summarize the Theory of Everything and wrestle it into a PowerPoint presentation that we could show to the scientists at the Albuquerque conference. First, though, we had to complete a whole bunch of calculations.

We'd already calculated the value of alpha, the fine-structure constant, at Samovar two nights ago. But the Theory of Everything also determined the values of 25 other constants, including the masses of all the fundamental particles—the electron, the muon, the tau, the Higgs, the three kinds of neutrinos, the six kinds of quarks. To prove beyond a doubt that the theory was correct, I wanted to use its equations to calculate *all* those constants to a high degree of precision, at least ten significant digits for each value. So I did the math as we cruised across the Midwest, and Andrei double-checked my work, scribbling equations just a couple of feet to my right. Luckily, Marguerite let us use her laptop, which was equipped with a calculator and the PowerPoint software.

It took a lot of effort, but I was glad to be busy. Andrei and I had so much work to do, we couldn't talk about anything else. In particular, we avoided any mention of the argument we'd had last night, and that was fine with me. I didn't want to think about God right now. Mathematics was more than enough.

Behind us, Marguerite's brother sprawled full-length across the minivan's third row of seats. Ben had his own laptop, and he was playing his favorite game on it—Battleground—a loud, violent combat simulation that he never seemed to get tired of. His game controller was connected by USB cable to the computer, and he wore a pair of headphones too, but the program was so noisy that I could still hear all the gunshots and explosions. Although Ben was clearly a jerk, he wasn't an idiot; I noticed that he'd made some custom modifications to his controller, gluing extra pieces to the device to improve his grip on it. He was a tinkerer, a guy with mechanical smarts. He'd also modified the minivan's electrical system, extending the power lines to the back of the car to make it easier to recharge his laptop.

The noise from Ben's war game was definitely annoying, but what really drove me nuts was Marguerite's humming. Over the past four hours, she'd hummed snatches of "Despacito," "Shape of You," "No Tears Left to Cry," and "Love Yourself." It was such a sappy, sickening collection of songs, and they sounded even worse when Marguerite droned their melodies from behind the steering wheel, the low notes thrumming at the back of her throat, the high notes vibrating in her sinuses. Just as I was finishing a particularly difficult calculation, she hummed the chorus of "Despacito" for the nine millionth time, and I felt like my brain was going to shatter. I couldn't take it anymore.

I leaned forward, angling toward the gap between the driver and passenger seats. "Uh, Marguerite? I hate to bother you, but that song is really distracting."

"What? What's that?" She sounded as if she were coming out of a trance. "A song?"

"Yeah, I'm trying to do some math, and your humming makes it hard for me to concentrate."

"Oh! This is embarrassing! I'm so sorry!" She let out a forced laugh and pointed at the highway ahead of us. "When I'm driving, I space out sometimes. I didn't even realize I was humming!"

I found this hard to believe, but I nodded anyway. "It usually doesn't bother me, but like I said—"

"No, no, I understand! You're working on your theory, right? The Theory of the Whole Universe?"

Charlotte looked up from the phone in her hands. "It's called the Theory of Everything." She glanced at me over her shoulder and pointed at the phone's screen. "A little while ago I went to a science website to read up on string theory. I wanted to understand why you were so excited about it."

I was touched. I closed my math notebook and gave Charlotte a big smile. "So, how far did you get? The extra dimensions, the manifolds? The vibrating strings?"

Her cheeks reddened a bit. It only made her more beautiful. "Well, I didn't understand a lot of the science, but I liked reading the historical parts. Did you know that Albert Einstein was the first scientist who tried to solve this problem? He wanted to discover how all the laws of physics were connected, but he couldn't figure it out."

Before I could respond, Marguerite nodded emphatically, mashing her dreadlocks against her seat's headrest. "That's because Albert Einstein didn't have God's help." To emphasize her point, she reached forward and slapped the dashboard. "The human race wasn't ready yet to see all of God's plans, so She didn't reveal them to Einstein."

I felt compelled to defend the greatest physicist of modern times. "Actually, Einstein did most of the groundwork for the theory. He put all the building blocks in place, the principle of relativity and the quantum picture of—"

"Okay, sure, but that stuff isn't so important now. God blessed *you* with the final revelation, and that's the one that counts." Although Marguerite kept her eyes on the road, she spoke as if she were addressing a congregation in front of her. I got the feeling that she saw something else through the windshield besides the highway, something the rest of us couldn't see. "You're gonna change the world, Joan. Everything's a mess right now because the human race is blundering through the darkness, but once we get to New Mexico you're gonna shine the light. You're gonna unveil God's holy design for the universe."

I heard a low unhappy grunt to my right. Andrei stopped scribbling equations in his notebook and gave me a pointed look. He said nothing, but I knew what he was thinking: *You've chosen the wrong side. You've foolishly abandoned the rational world of Science and joined forces with the insane crusaders of Religion.* The disappointment on his face was unmistakable.

Andrei was wrong, though. I hadn't changed my allegiances. I was still a mathematician, still a believer in science and logic, and I had no faith in Web-based spirituality. But I couldn't divulge my true feelings in front of Charlotte and Marguerite, who would've been horrified if I'd revealed how skeptical I was about the Church of Everlasting Light and all the other Internet religions. So I just sat there, my face burning.

Charlotte came to my rescue. She looked at me again over her shoulder. "Well, one thing's for sure—you're getting people's attention. I've sent messages to about thirty of my friends so far, telling them all about the Theory of Everything, and you wouldn't believe how fascinated they are. I mean, most of the Everlights are just regular kids who *hated* their science classes in high school, but now they're dying to know more about physics. Crazy, right?"

I beamed at her. This was exactly what I wanted to hear. "Maybe they're interested now because they realize how important it is." I gave Andrei a pointed look of my own. "When science and math are meaningful to people, they want to learn more about them. That's how it always worked for me, even when I was little."

Charlotte nodded. "And you know what else people are talking about? They're asking a million questions about your personal history." She pointed at me. "People want to know if you're Catholic and if your family is French. Because they're wondering if you're related to Joan of Arc."

I wasn't so wild about this piece of news. "Seriously? Because of my name? That's what they're focusing on?"

"It's a lot more than your name. You're getting messages from God, and that's what happened to Saint Joan. I read up on her history too." Charlotte raised the phone again. "If you want, I can show you the websites about her. She started hearing voices when she was a teenager, the voices of saints and angels. They came to her while she was in the fields with her family's flock of sheep. This was in the French town of Domrémy, back in the fifteenth century."

I was getting uncomfortable. I wanted to steer the conversation to a different subject, but I couldn't stop myself from pointing out an inconsistency. "Wait a second. She's 'Joan of Arc.' So didn't she come from the town of Arc?"

"No, that was a mix-up. Her family name was Darc, and some historian changed it to D'Arc. Then it was translated into English as 'of Arc.'"

Andrei let out another grunt. He'd closed his notebook so he could follow the conversation. "I wonder if the historians made other mistakes. Maybe other parts of the story are just as false."

Charlotte scrunched her eyebrows for a moment, probably wondering why Andrei seemed so irritated. Then she turned back to me. "Anyway, the saints ordered Joan to go to the king of France and take command of his army. They were in a war against the English, who'd invaded France and taken over the northern half of the country. It was called the Hundred Years' War because it literally went on for over a hundred years. And by the time Joan came along, the war was going really badly for the French. They were about to lose everything."

Marguerite slapped the dashboard again. "But no power on earth can stand up to God! That's how Joan of Arc defeated the English armies. She showed them the awesome strength of the Almighty and sent them running back to England!" She raised her chin and looked at me in the minivan's rearview mirror. "And you're gonna do the same thing in New Mexico, Joan. You're gonna lead an army of believers to that conference of atheists, and you're gonna show them God's power!"

I shook my head. This was why I'd felt so uncomfortable, why I'd wanted to change the subject. I didn't like this comparison with the other Joan, not one bit. "No, sorry, that's not how I see it. I'm going to the conference to make a scientific argument, a presentation about string theory. That's it. That's all I'm trying to do." I raised my notebook and waved it back and forth. "I'm not a soldier of God. I'm a mathematician."

Charlotte loosened her seatbelt and twisted around. She leaned through the gap between her seat and Marguerite's, trying to get as close to me as possible. She could see how upset I was, and she wanted to comfort me. "You're right, there's lots of differences between you and Joan of Arc. For one thing, you're a city girl. You've never herded any sheep, right? Or worn a suit of shining armor?" Smiling, she pointed at my jeans and T-shirt, the one that said RESIST in big black letters. "But you have to admit, there are similarities too. Joan of Arc was seventeen when she took charge of the French army, and you're seventeen. And that video of you from two nights ago? The one that shows you running down the street with the rainbow flag? Take a look at this."

She tapped the screen of Marguerite's phone, then held it up for me. It displayed a painting of a medieval battle, a jumble of knights and horses and swords, all clashing on a muddy, corpse-strewn field. At the center of the picture was a white horse rearing back on its hind legs, and mounted on the horse was a woman in silver armor, holding a flag in her right hand. It wasn't a rainbow flag, but it was brightly colored. It showed God in a red robe and a pair of kneeling angels.

The tip of Charlotte's finger touched the screen, brushing ever so lightly against the image of Saint Joan. "I mean, just look at it. Look at the flag she's holding. You can see why people might make the connection, right? Between you and her?"

She leaned farther through the gap, so I could see the phone better, and she stretched her other hand toward the backseat and rested it on my knee. My insides jolted as I stared at Charlotte's face, because it was so glowing and ecstatic, so full of yearning. But was she really yearning for *me*, Joan Cooper of West 78th Street? Or was she hungering for something else altogether? Did she want a girlfriend or a saint?

I frowned. I felt annoyed and confused. I wanted to get back to my calculations. "Yeah, sure, I can see how people might think of the similarities. But I don't want to encourage it, okay? I don't want people thinking I'm a saint, because I'm definitely *not*."

Marguerite looked at me in the rearview again. "Don't worry, Joan. The folks we've been getting in touch with online are very understanding. The last thing they want to do is make you uneasy. You'll see what I mean when you meet them tonight."

"Meet them?" I was so surprised, I actually jumped. My butt lifted half an inch off the backseat. "What are you talking about?"

"We're gonna stop tonight at the Granite City campground in southern Illinois. That's the meeting place we announced on Facebook and Snapchat and in all the messages we've been sending out." Marguerite pointed at the phone in Charlotte's hand. "Now, I don't want to get your hopes up too high. We've had bad luck in the past when we tried to arrange real-world meetings like this. People say they're gonna come, but most of them never do. They'd rather stay at home and hang out in their chat rooms. If we're lucky, though, maybe a few folks will show up."

Still disoriented, I looked at Charlotte. "Why didn't you tell me about this?"

She patted my knee, trying to soothe me. "We talked about it last night, remember? We said we'd try to round up some supporters to come with us to New Mexico. That way, you'll have a better shot at persuading the conference organizers to let you speak."

"But why are we stopping in Illinois? That's still a thousand miles away from Albuquerque. We should drive through the night and get there as fast as possible."

Marguerite shook her head. "First of all, this minivan wouldn't make it. We need to let the engine cool overnight. Second, we don't have to get to the conference till Sunday, so we'll have all of Saturday to drive to New Mexico. We can hit the road again right after the prayer meeting tomorrow morning. It's scheduled to start at dawn, so if you keep your speech short we'll be done by eight o'clock."

"Prayer meeting?" I raised my hands to my head and ran them through my hair. This was getting worse and worse. "Are you saying you've arranged a religious service? And you want me to be in it?"

"Yep, that's the plan." Marguerite shrugged, jostling her dreadlocks. "We're calling it 'A Sunrise Prayer for Peace,' so maybe it'll attract some of the people who are worried about Russia. And we're gonna pray at the Cahokia Mounds, which is a good spot for this kind of gathering. It's a state park, open to the public, no entrance fee. It's also a Native American archaeological site, so it has a huge concentration of spiritual energy. Plus, it's only ten miles from the campground."

I grabbed a fistful of my hair and tugged at it until I felt the sharp pain in my scalp. I couldn't believe I'd let them talk me into this. Charlotte saw my distress and kept patting my knee, but it didn't help at all. Andrei was no help either; he just opened his notebook and went back to his calculations.

And just when I thought things couldn't get worse, I heard a breathy, creepy voice behind me.

"You know how Joan of Arc died, right?"

I saw Marguerite's brother out of the corner of my eye. Ben had taken off his headphones and leaned all the way forward. His stubbly chin jutted over the back of my seat and his mouth was just a few inches from my right ear.

"Come on, Joan. If you're so smart, you should know the answer."

His voice startled me, but I stiffened instead of flinching. I was determined not to show any fear or discomfort, so I said nothing to the creep. I pretended that I hadn't seen or heard him.

But Charlotte saw him and pulled back, retreating to her seat. Marguerite glared at her brother in the rearview, and Andrei looked up from his notebook. He put down his pencil, very slowly and deliberately, and clenched his hands. It looked like he was getting ready to defend me.

In response, Ben laughed. "Hey, you're a Russkie, right? I can tell from your accent, so don't deny it, bro." He shifted farther to the right, edging closer to Andrei. "How long have you been in America, Russkie? And what are you gonna do once we start bombing the crap out of your country?"

"*Shut up, Ben!*" Marguerite's voice rang across the minivan. "*Stop bothering my friends!*"

Ben held up his hands in mock surrender. "Relax, sis. I'm just contributing to the mesmerizing discussion you got going here." He lowered his hands and slapped the back of my seat, trying to make me jump. "I mean, all this talk about Joan of Arc is very entertaining, but you girls left out the best part. Don't you remember the story from history class? How the English got their revenge on little Joan?"

"*No one cares what you think! Go back to your stupid computer game!*"

"No, I'm sorry, this needs to be said. Joan of Arc was captured during a battle. The English put her on trial and sentenced her to death." Ben shifted in his seat again, moving away from Andrei and back to me. "They tied her to a stake in the middle of the town square and dumped a big pile of dry wood at her feet. Then they set it on fire and stepped back to watch the show."

"*Stop it! You're scaring her!*"

It was true. I was scared, more scared than I really should've been. Ben was a lowlife, a small-minded jerk who antagonized everyone around him, and I should've just ignored everything he said. But he was getting to me. I stared straight ahead, trying not to look at the guy, but I could feel his breath on the right side of my face. It smelled like sour milk.

"Joan of Arc started screaming. First she screamed for water. Then she screamed for Jesus. Then the flames and smoke covered her, and the screaming stopped." Ben's voice rose. He was enjoying this. "The fire burned her clothes right off her. When the smoke cleared, she was as black as an overcooked steak."

"*You're a pig, Ben! You're subhuman, you know that?*"

It wasn't like me to just sit there and take it. Under any other circumstances, I would've spun around in my seat and told Ben what he could do with himself. But by this point I was so frightened that my skin had gone cold. I felt like I was going to throw up. And it wasn't because he was a really ghoulish storyteller or anything like that. What terrified me was the fact that Ben seemed to know something about me that no one else in the minivan knew. I got the feeling that he'd done some research on the Internet and discovered my worst secret. And now he was using it to torture me.

"But the English weren't finished with Joan yet. They knew the French would see Joan as a martyr, a saint who died for her country. And during the Middles Ages the Catholics used to preserve the corpses of their saints, or whatever body parts they could find. It's weird, but they liked to worship the bones and hair and teeth. You know, the 'holy relics.'" Ben lowered his voice, getting ready for his story's climax. "Anyway, the English didn't want that to happen. They wanted to destroy every last trace of Joan. So they poured oil over her corpse and burned it again. And when there was nothing left but her charred skeleton, they got their hammers and smashed her bones to dust."

I buried my face in my hands and started crying. This was what Ben had been hoping for, the payoff for all his goading, and I hated to give him the satisfaction of seeing it. But I couldn't stop myself.

He knew what had happened to Samantha. He must've learned my full name, probably by sneaking a look at my math notebook. Then he'd put my name in a Web search and found the news articles describing how my sister had died. Now he was taunting me with the story of Joan of Arc's execution because he knew Samantha had burned to death.

And somehow he'd also guessed that it would happen to me.

* * *

The Granite City campground was neither woodsy nor picturesque. It was located in an industrial section of southern Illinois, less than half a mile from the highway, and it was bordered on three sides by immense parking lots for trucks.

The sun had just set by the time we drove into the campground. Marguerite parked in front of the registration office, and we all clambered out of the minivan to stretch our legs and take a look around. There were several small cabins and at least thirty recreational vehicles packed into the dusty, bare-dirt lot, along with a dozen brightly colored tents. I was a little surprised that the place was so crowded, since the summer vacation season was long over. Also, the campground wasn't that attractive to begin with.

Then I heard someone in the distance shout, "They're here!" Within seconds, a hundred people streamed toward us.

They emerged from the cabins and tents and RV's. Most of them were young people, in their late teens and early twenties, and most were dressed like hippies or New Agers, in patchwork pants and fringe jackets and tie-dye shirts and headbands. The vast majority had tattoos on their arms—butterflies, flowers, mandalas, sunbursts—and quite a few also had peace signs painted on their faces. And all of them were smiling as they hurried across the campground, talking in excited voices, their eyes fixed on me.

I retreated toward the minivan until my back pressed against the driver side door. Everyone in the crowd seemed to recognize me, and though I knew the reason why—they'd probably seen me in the videos that Charlotte had spread to every corner of the Internet—it still freaked me out. The people at the campground were so exultant, so happy to see me. It was too much. I couldn't handle it.

Marguerite was also overwhelmed. She dropped to her knees in front of the minivan and raised her arms toward heaven. "Oh thank you, Lord! I expected only a handful, and you delivered a multitude! Thank you, *thank you!*"

When they reached us, the people surrounded Marguerite and grasped her hands and helped her to her feet. Then they formed a circle around the Honda Odyssey, and a hush fell over the crowd. The people at the front kept a respectful distance from the minivan, but they were clearly eager to close in and embrace me.

Marguerite spun around, gazing in pure bliss at everyone there. She swept her arms in a big arc, taking in the entire crowd.

"Here they are, Joan!" Her voice boomed across the campground. "Here's your army!"

Chapter Twenty-Two

I didn't get much sleep that night. Which partly explains why the next morning was such a disaster. I should've seen it coming.

I slept in one of the campground's cabins. I locked the cabin's door and spent the whole night alone. Marguerite and Charlotte mingled with the other people at the campground, making friends with all the seekers and believers whom they recognized from Snapchat threads and Reddit forums, but I couldn't bear to face them. So I just sat on the cabin's creaky bed and worked on my calculations. I filled the last pages of my notebook with equations, using string theory to enumerate the fundamental constants of the universe.

At 10 p.m. Andrei knocked on the cabin's door. He delivered my dinner—fried chicken from a takeout place close to the campground—and offered to come inside and double-check my math. But I said no. I closed the door and locked it again and ate my dinner. Then I opened Marguerite's laptop and worked on my PowerPoint presentation. I started composing the paragraphs that would summarize the Theory of Everything for the scientific world.

I made good progress. There was no one to interrupt or distract me, and by 2 a.m. the presentation was half finished. But I had another reason for wanting to be alone that night, and it had nothing to do with science or PowerPoint.

I was hoping for another visitation. I wanted to talk with God again, even if only for a few minutes. I felt so lost, so uncertain about what to do next. I had no idea how to deal with all the strangers who'd just joined us, the hundred followers from all over the Midwest who'd responded to our call and gathered at the campground and seemed so excited to meet me. And I thought that if I stayed alone in that cabin, I'd hear another knock on the door, but this time it wouldn't be Andrei. It would be the Almighty Herself, in the form of a young girl or an old man or a talking camel or whatever. And She would step inside and tell me everything I needed to know.

At 3 a.m. I turned off the lights and tried to sleep, but it was a struggle. I woke up every fifteen minutes, convinced that God stood right outside the cabin. But when I finally heard a knock on the door, it was only Marguerite. She said it was five-thirty, time for us to go to Cahokia Mounds State Park. It was cold outside and the sky was still pitch-black, but the sun would rise in an hour and a half, and we needed to prepare for the prayer meeting.

I got out of bed and brushed my teeth with the complimentary toothbrush I found in the cabin's bathroom. Then I grabbed my math notebook.

By 6 a.m. all of us were back in the minivan—Marguerite, Charlotte, Andrei, Ben, and me. But now we rode at the front of a caravan of RV's. We led our followers to Route 111, and after fifteen minutes we turned into a parking lot off Collinsville Road. As we got out of the car I noticed that the eastern half of the sky was brightening. The western half was much darker, but I could make out an odd shape on the horizon, a broad hill with a flattened top, about a hundred feet tall and a thousand feet across. Charlotte saw it too and grasped my arm. She let out a delighted yelp and pulled me forward.

"Come on! That's it!" She pointed at the hill. "That's Monks Mound!"

"What?" I was still thinking of Joan of Arc and the Middle Ages, and for a moment I thought we were headed for a medieval-style monastery. "There are monks up there?"

She shook her head. "Nah, that's just the name the white settlers gave it when they came here. The mound was built by Native Americans a long time before that." She held up Marguerite's phone, our connection to the Internet. "The park's website says it's a thousand years old."

I squinted at the mound, trying to glimpse its outline against the gray sky. In geometric terms, it was shaped like an acute trapezoid. I also saw a few smaller mounds to the south. "There's more of them over that way."

"Yeah, there used to be a city here, the capital of a great civilization. The archaeologists call it Cahokia, but the Native Americans who lived here didn't have a written language, so we'll never know the city's real name. They built hundreds of earthen structures in the area, but Monks Mound is the biggest."

Marguerite hung back in the parking lot to direct the RV drivers—there wasn't enough room for all their vehicles—and Ben loudly announced that he had no interest whatsoever in the prayer meeting, so he was going to stay behind in the minivan and play another game of Battleground on his laptop. But Andrei followed Charlotte and me down the path that led to the base of the mound. At the end of the path was a modern stairway built into the mound's side, an easy way for the tourists to get to the top. Charlotte leapt up the steps, still pulling me along.

"This is amazing! Can you feel how special this place is?" She squeezed my arm. "This mound was the spiritual center for thousands of people, a whole civilization. And it's still vibrating with all their energy!"

To be honest, I didn't sense any Native American vibrations, but I definitely felt Charlotte's energy. Her breath came fast as she climbed the stairway, her sandals slapping each concrete step, her yellow dress swishing around her legs. She got winded after half a minute—she wasn't in tiptop shape like I was—so I took the lead and grasped *her* arm and helped her up the stairs. I gripped her gently, a couple of inches above her elbow. I could feel the blood pulsing just beneath her skin.

By this point the sky had turned a vivid shade of purple. As we climbed higher I saw the flat floodplain all around us, etched with levees and canals and horseshoe-shaped lakes. I could also see downtown St. Louis on the western horizon, its skyscrapers and Gateway Arch on the other side of the Mississippi River. I felt dizzy for a second, but not because of the altitude; we were only a hundred feet above the plain, and that isn't very high for someone from Manhattan. No, I felt disoriented because I was happy. I adjusted my grip on Charlotte's arm and she edged closer. I was convinced now that she really *liked* me, liked me in the same way that I liked her.

Then we reached the top of the stairway, and a cold wind slapped my body. At least three hundred people stood on the grass at the flattened top of the mound, scattered across a plateau about the size of a football field. I couldn't see the people so well—it was still pretty dark—but I could tell they were all facing me. They just stood there, motionless and silent.

I let go of Charlotte and took a step backward. Then another. I probably would've fallen back down the stairway if Andrei hadn't come up behind me and placed a hand on my shoulder. Charlotte curled her arm around my waist and pulled me away from the steps. "Well, look at this," she said. "I guess a few folks got here early."

It was much more than "a few." Hundreds of people were waiting in the cold darkness. Waiting for *me*. That was the scary part.

The wind gusted over the mound. I shivered and tightened my grip on my math notebook, bending its cover. "This is insane. Who are all these people?"

Charlotte shrugged. "I'm not sure. We put posts on Facebook and Twitter about the meeting, and a lot of people shared them. And everyone is freaking out over the Russia situation. Maybe that's why they're here."

"Is there more news this morning? Did you see anything on Marguerite's phone?"

"I only had time to read the headlines, but they weren't good. The Russian army is heading for Poland now. And the White House issued an ultimatum."

The wind blew harder, thumping my eardrums. We stood on the highest point for miles around, which explained why it was so breezy up there, and yet I couldn't shake the feeling that something else was going on. It felt like someone was trying to get my attention, like the wind was delivering a message from the Almighty: *Step up, Joan. The people are frightened. They need your help.*

Charlotte pointed at the other end of the plateau, a hundred yards away. "Marguerite asked one of her friends to bring a microphone and some loudspeakers. It looks like they're over there."

I gazed in that direction and saw a couple of big rectangular speakers set up on the grass. The sight made me shiver, but I shook off the fear. "All right. I guess I'm ready."

Side by side, we walked across the top of Monks Mound, with Andrei right behind us. The people in the crowd clearly recognized me; they turned their heads as we walked past, trying to keep me in view. I got a better look at their faces and noticed that this crowd was older and less colorful than the hippies we'd met at the campground. Most of the people here were ordinary, middle-class men and women, in jeans and windbreakers. Some were middle-aged, some were elderly, some had even brought their kids. I imagined them in their homes in St. Louis and Granite City, sitting in front of their TV sets and watching the news from Eastern Europe. Then I pictured them turning on their phones and seeing the post about the Sunrise Prayer for Peace. Maybe they'd wanted to do something positive and hopeful instead of just waiting for the next news bulletin. That way, they wouldn't feel so powerless.

When we reached the patch of grass where the loudspeakers were, we turned around to face the crowd. We stood in a line, Charlotte to my left and Andrei to my right. The sky to the east had turned bright orange, and sunrise was only twenty minutes away. The folks who'd traveled with us from the campground were now coming up the stairway in a steady stream and spreading across the top of the mound. I glimpsed Marguerite among them, shouting instructions. More and more people joined the crowd, and soon I realized that hundreds of latecomers were arriving at the state park. I saw them in the distance, parking their cars anywhere they could and dashing toward Monks Mound, hurrying to get here before the prayers began.

Then I spotted the TV news crew. A woman in a red pants suit appeared at the top of the stairway, followed by a man wearing a black windbreaker and lugging a video camera on his shoulder. The man's jacket was embossed with the letters KMOV, which I assumed was the call sign of a local TV station. The woman stretched her arm in our direction, and then she and the cameraman made a beeline for us.

Andrei saw them too. He leaned toward me, bringing his lips close to my ear. "This is a problem. We can't appear on the news." He gripped my arm. "We need to leave. Right now."

I shook my head. "No. We can't." An idea had just occurred to me. "Maybe it's a good thing that the news people showed up. Maybe they can help us."

"Joan? Don't you see the danger? The Russian spy agencies monitor all American television broadcasts. If we appear in a news video, their facial-recognition software will identify us." He tugged my arm, trying to pull me to the right. "Then Vlad will send a team of agents here. To kill us. Or he'll do it himself."

"But we'll be gone by then." I stood my ground, refusing to be pulled away. "We're gonna hit the road right after the meeting, remember?"

"Look, it's bad enough that videos of you are circulating all over the Internet, but if you go on national television you'll become so—"

He was interrupted by Marguerite, who ran over to us and whispered, "Here we go! Are you ready, Joan? Charlotte?" Then, without waiting for an answer, she picked up the wireless microphone that lay on the grass and turned it on.

"*Good morning, beautiful people!*"

Marguerite raised one arm straight up and beamed at the crowd. She slowly turned from left to right, sweeping her joyous gaze across everyone there, her dreadlocks swaying behind her shoulders. After a moment of hesitation, the crowd responded to her greeting, shouting "Good morning!" back at her in a thousand eager voices. The words echoed over Monks Mound, musical and booming, like the chorus of a pop song at maximum volume.

Marguerite nodded and her smile grew wider. Then she stretched her arm to the left and pointed at the eastern horizon. Just above the distant tree-covered bluffs, the sky glowed a brilliant yellow, brightening each second, just ten minutes from sunrise.

"Look at the dawn! What a magnificent sky! What a gift just to see it, what a miracle!"

The crowd murmured its assent. Someone started clapping, then a dozen more people joined in, and soon everyone on the mound was applauding the dawn. Marguerite kept smiling, tilting her head back as they cheered, and I realized she was really good at this. She had a knack for encouraging public displays of emotion, maybe because she was so sincere and enthusiastic herself. It was impressive.

She patiently waited out the applause, then lowered her head back to the microphone. "Thank you so much for coming to this sacred place, this ancient monument. Thank you for your wonderful kindness and generosity. And thank you most of all for your undying hope. Even in the darkest of times, even when war and catastrophe are on the horizon, there's always room for hope in our hearts. And hope is what will save us from the terrible crisis we're facing now!"

The people cheered again, even louder this time. Some of the New Agers from the campground had zigzagged their way to the front of the crowd, and they applauded wildly just a few yards away from us. A red-haired teenage girl wearing sunglasses and a green dashiki jumped up and down on the grass. Standing next to her was a goateed dude in a threadbare denim jacket, and next to him was a tall, gray-haired woman with a crown of daisies on her head and a canvas bag slung over her shoulder. And at the very center of the front row was the TV cameraman, who knelt on the grass and pointed his shoulder cam at Marguerite's ecstatic face.

"I want to tell you what's giving *me* hope. A few days ago I heard about a young woman who lived in New York City. This woman was just seventeen, just a high-school senior, and yet she was blessed with a spectacular mind that could solve any mathematical puzzle. Even more spectacular, though, were this young woman's visions. She claimed that God had spoken to her and revealed a great secret, the mathematical design of the universe. And God commanded her to share this scientific theory with the whole world, because a horrible cataclysm was coming, and only the revelation of God's design could stop the disaster."

As Marguerite spoke into the microphone, she stepped backward and turned sideways, positioning herself next to me. The crowd shifted its collective gaze, and I held my breath as everyone looked me up and down, scrutinizing my rumpled jeans, my grungy T-shirt, my flyaway hair. I raised my math notebook to my chest, holding it like a shield. I didn't want their attention. I wanted to turn around and bolt.

But Marguerite wrapped her arm around my shoulders, steadying me, anchoring me. "Her name is Joan, my friends. And here's the amazing thing: just an hour after she told me that a disaster was going to happen, I saw the news from Eastern Europe, and it confirmed everything she'd said. So, yes, there's something we can do to stop this war and save our world from destruction. We have to help Joan complete her God-given task." She squeezed my shoulder. "But I've said enough. Why don't we let Joan tell the rest of the story?"

Then she handed me the microphone. She stepped to the side, leaving me alone to face the crowd.

The people were silent. I kept waiting for them to cheer again, but everyone just stood there, staring. The gray-haired woman frowned and the teenage redhead furrowed her brow. The goateed dude folded his arms across his chest.

I could guess the reason for their change of mood. They'd come here expecting a more conventional prayer meeting—you know, prayers for world peace, speeches about the horrors of war, maybe a performance of a John Lennon song or a release of white doves. But instead of all that, we'd given them a story about math and science and a divine mission, and now they were confused. And frankly I didn't blame them. How could a scientific theory stop a war? It didn't make sense to me either.

But that was my task. That's what God told me to do. And I trusted Her.

I raised the microphone to my mouth. "Uh, hello. I'm Joan. I'm here to talk about string theory."

No response. I thought I might get a few chuckles—what I'd just said was so absurd—but the crowd remained silent. The gray-haired woman grimaced.

"Okay, I realize this is supposed to be a prayer meeting. But until a few days ago, I wasn't a religious person. So I don't know any prayers. Sorry about that."

I stopped and lowered the microphone, at a loss. The TV guy was pointing his camera at *me* now, which was pretty nerve-racking. I glanced to the right, looking to Andrei for some help, but he just shook his head. Marguerite was clearly anxious—she gave me a frantic look and twirled her right hand in a "hurry up" gesture.

Charlotte, though, just smiled. She didn't seem worried at all, and that gave me the confidence to continue.

"But maybe prayer shouldn't be our top priority right now. I mean, when you pray, you're usually asking God for something, right? In this situation, though, the opposite thing happened—God asked *me* to do something. And maybe that's the way the universe is supposed to work. Maybe God provides the advice and guidance, but we're the ones who have to do the heavy lifting, you know?"

I paused again, waiting to see if anyone agreed with me. I didn't get any applause, not even a smattering, but the redhead nodded and the goateed dude let out a grunt. So they were listening at least.

"Look, you don't have to believe my story. Some people think I hallucinated my conversations with God. Actually, *most* people think that. But *I* believe it, so I'm gonna follow God's instructions." I nodded firmly, decisively. "Right after this meeting, I'm going to Albuquerque, New Mexico, where they're holding a conference on science and religion. I'm gonna present a solution to string theory, a unique set of equations that describe the shape of our universe and everything in it. It's a pretty complex theory, so I won't go into the details. But I'll tell you one thing: It's the most beautiful piece of mathematics I've ever seen." I turned left, toward the east, my eyes drawn toward the blazing horizon. The sun was going to come up any minute now. "The theory is glorious. That's the only word that comes close. I don't know if it's beautiful enough to convince people to stop fighting each other. But I think there's a chance."

I glanced to the right again. Andrei was still doubtful, still shaking his head, but Marguerite's face gleamed in the morning light. She clearly liked what I was saying, and the crowd seemed to be warming to me too. The people nodded and stared at me as the wind whistled overhead. I imagined they were starting to understand, starting to see the connections between mathematics and peace.

"I'm facing a big obstacle, though. I'm only seventeen, and I don't have a Ph.D. The scientists and philosophers at the conference have no idea who I am, and they're gonna find it hard to believe that a high-schooler discovered the Theory of Everything. And I can't send the equations to them by email or anything like that, because the theory can be used for evil purposes as well as good, and I want to make sure that the responsible people see it before the bad guys do. So it's a problem. How can I convince the scientists to let me speak at their conference?"

The question hung in the air, echoing over Monks Mound. I didn't expect anyone in the crowd to answer it, and no one did. But that was all right. I already knew the answer.

"Okay, I'll cut the suspense. I have a plan. You see, there's a bunch of numbers called the fundamental constants. They're like the manufacturer's specifications for our universe—they specify how fast it's expanding, how heavy are the particles, how strong are the forces. Scientists have done experiments to measure all these quantities, but their results provide only *approximations* of the constants rather than the exact values, because even the best experiments have small measurement errors." I stopped for a second to let this fact sink in. "But I can use the Theory of Everything to calculate the *exact* values of the constants. And that gives me a way to convince the scientists in New Mexico to listen to me."

I looked at the TV cameraman kneeling in the front row. I stared right into the lens of the video cam on his shoulder. My plan depended on him capturing these words, this moment.

"I want to focus on alpha, the fine-structure constant. It specifies the strength of the electromagnetic force in the present-day universe. This was the first constant I calculated, so I went on the Web to check my results, and I learned that some scientists in California recently did an experiment to measure alpha with greater accuracy than ever before. They made the world's most precise measurement of the constant, but they're still writing their research paper, so they haven't publicized their work yet." I lowered the microphone for a moment so I could open my math notebook to the right page. "Now I'm gonna read the exact value of alpha. I'm hoping the news media will record this number and spread it across the Internet. And when those scientists in

California see that my value matches their best-ever measurement, they'll realize that I've truly discovered the Theory of Everything."

The crowd went silent again, but this time it was the tense silence of anticipation. The redhead and the goateed dude leaned forward slightly, bending toward me, and the gray-haired woman gripped her shoulder bag. And at just that moment, the sun peeked over the horizon. Its brilliant rays skimmed the distant wooded bluffs and threw golden light on everyone standing on Monks Mound.

"Here's the exact value of alpha, the magic number of the universe, calculated to thirteen significant digits." I took a deep breath and looked down at the page, now tinged yellow from the sunrise. Then I recited the number into the microphone, pronouncing each digit slowly and carefully. "Zero…point…zero…zero…seven…two…nine…seven…three…five…two…five…seven…one…two…nine…three."

By the time I finished, the sun had climbed fully over the horizon. The redhead clapped in delight, and pretty soon the whole crowd was applauding. I wasn't sure exactly what they were cheering—the sunrise? The Theory of Everything? The mathematical proof of the existence of God? Or maybe they were cheering all those things. Maybe they were celebrating the fact that they hadn't surrendered to despair or lost the strength to hope and pray.

It was a wonderful moment, totally transcendent. I smiled at the crowd and closed the notebook.

In that same instant, the gray-haired woman stepped toward the TV guy and grabbed his camera. Before he could react, she flung it off his shoulder, smashing it to pieces on the ground. Then she reached into her bag and pulled out a gun.

I stood there, frozen, as she pointed the pistol at me. I had no time to wonder who she was. My only thought was a burst of confusion. *Is this the end? Is this what God wants? Or did I fail Her?*

Then the gunshot boomed over Monks Mound.

But while I waited for the bullet, someone dove in front of me. A body leapt diagonally from the right, knocking me sideways and backward.

I landed on the grass, jangled but still breathing. I tried to lift my head, and everything whirled around me—the sky, the rising sun, the hundreds of stampeding feet and terrified faces. I saw Andrei standing motionless a couple of yards away, just as paralyzed as I'd been a second ago. I saw Charlotte stepping backward, screaming in horror, and I saw Marguerite running across the mound, joining the stampede toward the stairway.

I didn't see the gray-haired woman. She must've run off after firing her gun. But when I sat up I saw the man who'd jumped in front of me, lying on the grass a few feet to my left. It was the goateed dude in the ragged denim jacket. He lay on his side, conscious but very pale. The T-shirt under his jacket was soaked with blood.

I crawled over to him. "Where were you hit? Let me see if I can—"

"I'm all right." His voice was calm. He actually grinned at me. "I don't feel any pain."

"You're going into shock!" I reached for his shirt, trying to locate the wound. "You need to—"

"What I meant was, I *never* feel pain." The sunlight struck his face, restoring some of its color. His eyes glinted gold and blue. "I can take the shape of a dying creature, but I can't experience its agony. Because I can't die."

I pulled my hand back. It was God. I didn't want to touch Him.

He coughed, spraying blood on the grass. "You need to leave before the police get here, Joan. Go back to the minivan and drive straight to New Mexico."

"But what about that woman, the one who shot—"

"*Just go!*"

I staggered to my feet. Fighting off the dizziness, I picked up my notebook from the grass and looked around for my friends. Within seconds, I found Andrei and Charlotte, and then we sprinted across the mound and back to the parking lot. Marguerite was already in the driver's seat of the minivan, yelling at us to hurry up. Ben was in the backseat, with his headphones on, still staring at the game on his laptop.

We peeled out of the state park and raced toward the highway.

Chapter Twenty-Three

Just before midnight we stopped at a 24-hour diner in New Mexico, at exit 333 off Interstate 40. It was near Tucumcari, a smallish town at the edge of a vast desert, 175 miles from Albuquerque.

The diner was ugly and the food was terrible, but we didn't care. We were too exhausted. I sat next to Charlotte at one of the diner's booths, and Marguerite and Andrei sat at the other side of the table. Ben didn't eat with us; he chose a different booth, propped his laptop on the table, and devoured a cheeseburger while staring at the screen. He didn't have the game controller in his hands, so I assumed he was taking a break from Battleground.

Ben had been playing that stupid game almost nonstop ever since we left Cahokia. He'd been so immersed in it while we sped away from the state park that I wondered at first whether he'd even noticed that something had gone wrong. But an hour later, after we crossed into Missouri, he checked his phone and informed us that the first news reports of the shooting had appeared on the Web.

I refused to look at the news. I was in denial, I guess. All through the morning and afternoon, while Marguerite drove the overheating minivan across Missouri and Oklahoma, I suppressed all thoughts about the shooting and focused on my PowerPoint presentation instead. I finished the summary of the Theory of Everything, putting the final touches on the file in Marguerite's laptop, while Andrei silently double-checked my math. Despite the minivan's mechanical troubles, we reached the Texas Panhandle by dusk, driving a steady 60 mph across the darkening plains, letting most of the traffic pass us. Cell-phone service was spotty, but the signal strengthened after we cruised into New Mexico. When we got to the diner we agreed to check out the latest news bulletins as soon as we finished eating.

But after dinner was over and the restaurant's lone waitress cleared our dirty plates, I was still apprehensive. The place was practically empty—there were only three other customers aside from us, and they all sat far away—but I still craned my neck and looked around, worried about eavesdroppers and spies and assassins. Marguerite handed her phone to Charlotte, who knew all the best news websites, and she quickly found an updated story on St. Louis Today dot-com.

She squinted at the phone's screen, bunching her blond eyebrows. "All right, the headline is 'Fatal Shooting at Prayer Vigil.' Hold on, the text is still loading."

Marguerite leaned across the table. "Is it slow? I really need to get a new phone."

"No, no, it's coming up now. Okay, here's the story." Charlotte brought the phone closer to her eyes. "'Illinois police are searching for the perpetrator of a fatal shooting this morning at a prayer vigil held at Cahokia Mounds State Historic Site. The victim, a man whom police are still trying to identify, was one of hundreds of people who attended the vigil, which was advertised on social-media websites as a Sunrise Prayer for Peace. The attendees said the religious ceremony was a response to the outbreak of armed conflict in Eastern Europe.'" She cocked her head. "Huh, that's weird. Why can't they identify the guy who got shot? He wasn't carrying a wallet?"

I hadn't told them who the victim was. I was still too shocked to talk about it. I guess I shouldn't have been so surprised that the Almighty would take the form of a dying man. Wasn't that the whole premise of Christianity? But reading about it in the Bible was very different from actually *seeing* it. I couldn't shake the memory, the image of the pale, goateed man coughing blood on the grass.

Andrei, who'd been very quiet during dinner, looked up from the table. "Perhaps the man was homeless? I saw him there, on Monks Mound, standing in the front row. He seemed a little strange."

I thought of what God had told me when I saw Her on the subway car. *I'm not homeless. The whole world is my home.* Then I pointed at the phone in Charlotte's hand. "What about the shooter? What does the story say about her?"

"Wait, it's in the next sentence. 'According to witnesses, a woman in the crowd fired a pistol at one of the speakers at the ceremony, but the bullet hit a bystander instead. The suspect, who fled the scene, was described as tall, middle-aged, and dressed in a counter-cultural style. She reportedly wore a crown of daisies in her hair.'" Charlotte made a face. "Great. Now the cops have another excuse to harass the hippies."

Andrei nodded. "It was a disguise, of course. She wanted to blend in with the crowd. But that particular disguise is a clue to her real identity." He tapped his head. "Russian women wear crowns of flowers at weddings. It takes some skill to make a good one, and her crown was very pretty."

It was a struggle to stay calm. I tried my best to keep my voice down. "So she was a Russian agent?"

"One of Vlad's, I'm sure. He must've intercepted the emails about the prayer meeting. And watched the videos of you on YouTube. Then he put two and two together and guessed that you would be there." Andrei gave me a pointed look. "I told you it would be dangerous, didn't I?"

Before I could respond, Charlotte clapped her hand to her forehead. "Oh my God, you need to hear the rest of this. 'Police are seeking to question the intended target of the shooting, a young woman who also fled the scene. Witnesses said she was a charismatic speaker, reportedly just seventeen years old, known only by her first name, Joan. Illinois police officials believe she may have important information about the incident, and they are urging law-enforcement authorities in other states to be on the lookout for her. A video of her speech at the Cahokia park, recorded by a KMOV-TV news team, has been distributed nationwide.'" Charlotte tapped the phone's screen. "And there's a link to a clip from the video."

She held up the phone for all of us to see. The video clip loaded after a couple of seconds, and then I saw myself on the screen, speaking into the microphone, the rising sun illuminating the left side of my face. "Here's the exact value of alpha, the magic number of the universe, calculated to thirteen significant digits. Zero…point…zero…zero…seven…two…nine…seven…three…five…two…five…seven…one…two…nine…three."

My stomach churned. There was a good chance that my parents would see this clip. Dad might come across it while surfing the Web, or one of Mom's friends might tell her the news. And it would terrify them.

On the other hand, there was also a chance that the scientific community would notice the video. Someone might forward it to the physicists in California who'd measured the value of alpha. Once they saw that I already knew their extremely precise measurement, which they hadn't made public yet, they'd probably get curious. Maybe they'd watch the rest of the video and listen to what I'd said about my solution to string theory. And maybe, just maybe, they'd spread the word.

It was a long shot. But the Creator of the Universe seemed to enjoy improbabilities. I sensed Her invisible hand behind all these events, nudging people toward their assigned tasks, pushing them to perform acts of kindness and courage. She was doing everything She could to rescue Her world.

Andrei, Marguerite, and Charlotte stared at me. They were all waiting for my reaction to the video, but I didn't want to talk about it. Instead, I pointed at the phone again. "Let's check the other news. What's the latest from Europe? What's happening with the Russian invasion?"

Charlotte nodded. "Sure, sure. Maybe there's some good news for a change, right?" She tapped the phone's screen once more. "Okay, the big headline is, 'Russia Rejects U.S. Ultimatum.' Uh-oh, that doesn't sound so good."

"Go on, keep reading."

"There's a smaller headline below the big one. It says, 'Russian President Claims the Baltic States "Invited" Russian Forces to Occupy Lithuania, Latvia, and Estonia, and the U.S. Has No Right to Demand Their Withdrawal. The White House Reacts With Disbelief and Anger.'" Charlotte paused to absorb the news, which wasn't good at all. "Oh man. Should I keep reading?"

Marguerite shook her head. "No. It's too depressing."

"It's worse than depressing. It's freakin' horrifying." Charlotte lowered the phone and turned it off. "I don't get it. Don't the Russians realize they've gone too far? I mean, *we* have a crazy president too. Any minute now he's gonna get fed up and launch his nukes at them."

Andrei leaned back against the booth's vinyl backrest. He looked tired, defeated. "The Russians think they have a secret weapon, a software program that could short-circuit the American military's computers. Whether the software will actually work, I have no idea. But they're confident enough to take a gamble."

No one said anything else for a while. In the heavy silence, the diner's waitress came by our table, asked if we wanted dessert (no), and left the check in front of Andrei. (Typical! Why do they always assume the guy is paying?) I grabbed the check and stood up, but before I left the table I gave all three of my friends a resolute look.

"Listen to me. We're gamblers too. That's why we're going to Albuquerque. We're betting everything on God and string theory."

Then I headed for the cash register. On the way over there, I picked up Ben's check as well. He didn't even look up from his laptop.

I used my last $60 to pay for dinner. Andrei, Charlotte, and Marguerite followed me out of the restaurant, and Ben came along too after Marguerite yelled at him. As we crossed the parking lot, which was empty except for our minivan and a couple of tractor-trailers, Andrei sidled next to me and spoke in a whisper. "More agents will be coming, you know. Vlad's people are probably in Albuquerque right now, waiting for you at the convention center."

I didn't look at him. Instead, I looked up at the immense desert sky, sprinkled with thousands of stars. It was beautiful. "Like I said, it's a gamble. We'll have to be fast and smart. And lucky."

"I'm sorry, Joan, but I don't think that's very—"

A siren cut him off. I turned toward the noise and saw a New Mexico State Police car turn into the parking lot, its lights flashing. The vehicle careened past the tractor-trailers and screeched to a halt next to our minivan.

We all stopped in our tracks.

A state trooper in a black uniform got out of the driver's seat. He was tall, broad-shouldered, square-jawed, and intimidating, his face dead white under the black visor of his officer's cap. As he stepped toward us, he grabbed a flashlight from his belt and shone it in our faces.

I felt an urge to run. The cops in Illinois were looking for me. They wanted me to come back east to answer their questions, and by now they'd probably sent alerts to every state police force in the U.S., including New Mexico's. I strained my eyes to the right and gauged the distance to the edge of the parking lot, wondering if I could flee into the darkness. But even if I could escape the state trooper, what would I do then? I was a good runner, but I couldn't jog all the way to Albuquerque.

After a few seconds, the trooper lowered his flashlight and pointed at our minivan. "Who's the driver of this vehicle?"

Marguerite took a tiny step forward. "Uh, that's me, sir. Is there a problem? I don't—"

"License and registration." He towered over her, squinting at her dreadlocks. He had thin, colorless lips and dark, unfriendly eyes.

She pulled a wallet out of her jeans and fumbled inside it. Her fingers trembled as she passed the laminated cards to the trooper. "We're from Pennsylvania. Well, my brother and I are from there, and my friends—"

"Where are you headed?" The trooper gave her documents a quick glance, then slipped them into his uniform's shirt pocket. "You going to Albuquerque?"

She nodded tentatively, uncertain whether this was a good or bad answer. "Yes, we've been driving west on I-40."

"So are you part of the caravan?" His voice rose on the last word, pronouncing it with suspicion.

"Uh, sorry, I don't—"

"We stopped a dozen out-of-state vehicles on I-40 tonight. All of them were packed with young kids just like you, all headed for Albuquerque. You know anything about that?"

Marguerite started crying. She was flustered and confused, and so was I. Although I'd told the crowd at Monks Mound about the conference in New Mexico, I hadn't expected anyone to follow us there, not after the shooting. But apparently some people had kept the faith and continued the pilgrimage. I couldn't believe it.

Then Ben stepped forward. He took off his headphones and pointed at me. "That's the girl you're looking for, officer. She's the one who gave that speech in Illinois, the one who caused all the trouble this morning." He raised his laptop. "It's all over the news, so you probably saw it. Her name is Joan Cooper. On the Web, everyone calls her Saint Joan of New York."

The trooper scowled at Ben, showing an instinctive disdain for the squealer. Then he pointed his flashlight directly at me. He stepped closer and studied my face.

"Yeah, you're the one, all right. I saw you on TV this afternoon before I started my shift." He looked me up and down, frowning when he saw the word RESIST on my T-shirt. "You were also on the radio, KCLV-FM. They played your whole speech on their Drive-Time show. You've had quite a day."

I tilted my head back so I could look the trooper in the eye. I was going to try to reason with him. "Sir, I haven't broken any laws. I was just exercising my First Amendment rights at the prayer meeting. Freedom of speech, freedom of religion."

He kept staring at my shirt. His frown deepened. "My understanding is that someone *did* violate the law at your meeting. Someone committed murder. That's a pretty serious violation."

"But I had nothing to do with that. I didn't shoot anyone. I was the *target*."

"Right, and that means the police in Illinois need your help. They need to ask you some questions to figure out who wanted to kill you. And you have a duty to cooperate with them, for the good of public safety."

"Okay, I get it. I'll talk to the police as soon as I can. But there's something else I have to do first."

He shook his head. "No, you and your friends are gonna come with me. I'm gonna call for another patrol car, and we're gonna take all of you to the trooper station in Santa Rosa. Then I'll get in touch with the Illinois State Police. Their detectives can ask you questions over the phone." He put his flashlight back on his belt, which also held a nightstick and a gun holster. "You might have to wait in the station till morning, maybe early afternoon. But after you talk with the detectives, you'll be free to go."

I felt like screaming at him, but I kept my cool and thought it over. Although the trooper had no legal right to detain me, we were hundreds of miles from civilization. Instead of following the command on my shirt (RESIST!), I needed to reach an accommodation with this man. "Sir, I'd like to ask you a personal question. Do you believe in God?"

Luckily, the question didn't offend him. He simply nodded. "I do."

"And has belief in God changed your life?"

He nodded again. "I go to church every Sunday. I've been a member of First Baptist in Fort Sumner for ten years." He looked at his watch. "Nine hours from now, I'll be there for the Sunday-morning service. That's assuming you don't delay me too much and I get off my shift on time."

"Okay. That's wonderful. I respect your belief. And I'm asking you to respect mine." I raised my hand and tapped my chest, right over my heart. "If you saw the news about me, you already know what I believe. God gave me a task. He showed me the Theory of Everything and commanded me to share it with the world. So I have to be at the Albuquerque Convention Center at nine o'clock tomorrow morning. It's vitally important."

"More important than catching a murderer?"

"Unfortunately, yes. Many, many lives are at stake."

"Yeah, that's what they said about you on the radio." He rested his hand on the hilt of his nightstick. "They said you talked about your theory at that prayer meeting in Illinois. Said you told everyone that this theory was the only thing that could stop World War Three."

His voice was skeptical but not scornful. There was some curiosity behind it, or at least a willingness to listen. This trooper was a believer. He took this kind of thing seriously.

I smiled at the man. "I'm just repeating what God told me. I'm following His instructions."

"Oh yeah? How can you be sure it's God and not just some voice in your head?"

There was no way to be sure. But at that moment I was. "I saw Him, sir. Clear as day."

"So this was a visitation? Like when God spoke to Moses?"

I didn't know the Bible nearly as well as the trooper did, but I nodded anyway. "Yes, it was God. He proved it when He showed me the correct theory."

The trooper stared at me. Then he leaned closer. "Now let me ask *you* a question." He narrowed his eyes. "What's so great about this theory you keep talking about?"

"Well, it's kind of like the Old Testament, you know? It explains the creation of the universe like the Bible does, but in the language of mathematics. And this explanation is the most fundamental and comprehensive of all the—"

"But God already gave us His scriptures, and that's supposed to be enough. Why do we need another revelation?"

I raised my hands in surrender. "Okay, I'll be honest. I just don't know."

"And if this task is so *vital*, why would God pick a teenager for it? Does that make any sense?"

"Absolutely not. And I told Him so. But I guess He has His reasons."

The trooper took a step backward but kept his eyes on me. I couldn't tell if he was angry or impressed. He looked straight at me for five full seconds.

Then he reached into his shirt pocket, pulled out Marguerite's license and registration, and handed them back to her. "Here you go. Observe the speed limit, please. If you drive straight to Albuquerque, you'll get there in two-and-a-half hours."

Marguerite dried her eyes and thanked him. Charlotte and Andrei thanked him too, and Ben put his headphones back on. And I let out an enormous sigh of relief. Thank God for the Baptists. Or at least for this one particular believer.

The trooper went back to his patrol car and opened the driver-side door, but at the last second he looked at me over his shoulder. "Joan? You should listen to that Drive-Time show on the radio. You know what they called your speech?"

"You mean the radio announcers?"

He nodded. "They called it, 'The Sermon on the Mound.'"

"Seriously?"

"Hey, I thought it was funny."

Then he got into his car and drove back to the highway.

Chapter Twenty-Four

We were cruising the streets of downtown Albuquerque at 3 a.m., looking for a parking spot near the convention center, when we spotted the encampment.

A row of tents had been set up along the edge of a parking lot that was jammed with dozens of RV's and campers.

It was my army. I recognized some of the tents and vehicles from the Granite City campground, but there were new ones here too. From my seat in the minivan I squinted at the lot, which was dimly illuminated by a distant streetlight. I saw no people there—no one standing or walking among the tents and RV's, nothing but long shadows stretching across the asphalt. Everyone was asleep.

Marguerite found the lot's entrance and quietly parked the minivan in a space next to a mid-size Winnebago. She cut the engine and turned off the headlights, but she didn't budge from her seat. She leaned forward and peered through the windshield. "Look at this. How did they all get here before us?"

Charlotte glanced at me over her shoulder and rolled her eyes. "It's not rocket science. We were driving below the speed limit and stopping all the time. We should've gotten here hours ago." She opened the passenger-side door. "I need to stretch my legs."

I followed her lead and got out of the car, taking the laptop and both of the math notebooks with me. It was chilly outside. Andrei looked up at the sky, which wasn't as impressive as the sky over Tucumcari but still a lot darker than what I was used to in New York. Ben took off his headphones and surveyed the place, turning left and right, clearly a bit twitchy.

We were all dog-tired but I think I had it the worst, because I'd hardly slept the night before. I stared longingly at the Winnebago we'd parked next to, a boxy vehicle with the word "Spirit" painted on its side. It was twenty-five feet long and probably had a couple of beds or sofa beds inside, maybe not so luxurious but certainly a lot more comfortable than the seats in the minivan. I felt an urge to knock on the RV's skinny door, wake up the slumbering occupants, and ask them if they had any extra room. In just six hours I had to confront the world's smartest physicists and convince them that I'd discovered the Theory of Everything, so I really needed to get some sleep before then.

But I didn't have to knock. As if answering my prayers, the Winnebago's door opened and a young woman stepped outside. I didn't recognize her at first because she wore a pink bathrobe now instead of sunglasses and a dashiki, but I remembered the distinctive color of her hair. It was the redhead I'd seen on Monks Mound, the teenager who'd stood at the front of the crowd next to the gray-haired Russian assassin and the goateed Almighty.

Marguerite recognized her too. "Emma!" She rushed over to the redhead and gave her a big hug. Then she turned back to me. "Joan, this is Emma Dowling. I met her at the campground last night, and she was at Cahokia this morning. She's an Everlight, just like me and Charlotte."

Emma approached me cautiously, her gaze lowered. For a moment I was afraid she was going to kneel on the asphalt, but instead she clasped my hand. “I’m so glad you’re safe, Joan. When that crazy woman pointed her gun at you…I didn’t…I just couldn’t…”

“Hey, hey, it’s all right.” I tried to sound reassuring. “I’m fine, see? Everything’s fine now.”

Ben stepped forward and stuck his ugly face in front of me. “Fine? Are you kidding? We’re sitting ducks here.” He pointed at the row of tents at the edge of the parking lot. “The last person who took a shot at you was disguised as one of your freakazoid friends, right? So there could be another Russkie sharp-shooter right over there, sitting in one of those tents, taking aim at us this very second.”

Emma’s eyes widened. She squeezed my hand and pulled me toward the door of her RV. “Joan, come inside! Everyone, get inside quick!”

I resisted at first, convinced that Ben was just stirring up our fears. But Andrei whispered, “He’s right. We’re easy targets out here.” So I let Emma lead me into the Winnebago, and everyone else piled in behind us.

Inside, the RV was cramped but cozy. Emma escorted me to a padded bench seat that curved around a rectangular dinette table. Marguerite and Charlotte leaned against the counter of a galley kitchen—it had a small sink, an undersized oven, and a mini-refrigerator—while Andrei and Ben stood beside the driver and passenger seats at the front of the RV. At the back of the vehicle were the bathroom and a queen-size bed half-hidden by a curtain.

I pulled my hand out of Emma’s and waved it in a circle, gesturing at the RV’s interior. “Are you alone here? You’re not traveling with anyone else?”

She gave me a sheepish look. “The thing is, I sort of borrowed the RV from my parents. I told them I was going camping with some friends in the Wisconsin Dells, and then I drove to Illinois instead.”

I kept my face blank, but inside I was hopeful. The Winnebago had more than one bed. Above the driver and passenger seats was a shelf-like overhead bed that looked pretty comfortable. And I got the feeling that Emma might be willing to share her space with us, at least for the next few hours. But before I could make the request, Ben stretched his arm over our heads and reached for the window above the bench seat.

“Listen, we gotta take this threat more seriously.” He pulled the shade down over the window, then hustled over to the galley. “To be honest, I don’t really care if the Russkies bump off your precious Saint Joan. But if they take another potshot at her, the bullet might hit me instead, and I’m not gonna let that happen.” He pulled the shade down over the window above the sink, then headed for the queen-size bed. “So we need to start thinking about defense.

One of us should go outside and stand guard. You know, keep watch over the RV from a distance."

Marguerite gaped at her brother. "What's gotten into you? For your information, this isn't a computer game, and you're not a real soldier. So you should stop playing—"

"Shut up, Margo." Ben leaned over the bed and pulled the shade down over the last window. "I'm not tired, so I'll stand guard for the rest of the night. I'll start shouting if I see anyone sneaking around the Winnebago. If we're lucky, all your hippie friends will rush out of their tents and scare off the Russkie."

No one responded right away, but after a few seconds Emma stood up and went to the galley. She opened one of the drawers below the sink and rummaged through it, tossing aside a tape measure, a package of batteries, and a screwdriver. Then she reached deep into the drawer and pulled out a gun. It was a revolver, old and black and scraped.

Emma held the gun gingerly, gripping it between her thumb and index finger, the muzzle pointed down. "It's my dad's. He used to be a cop."

She handed it to Ben, who grasped the revolver with both hands. "Whoa, a Colt Detective Special. This is a collector's piece. They stopped making these guns twenty years ago." He held it up to the light and examined the thing. Then he pulled a release switch, and the gun's cylinder swung out. "Fully loaded too. Very nice."

Needless to say, this was alarming. Ben seemed way too familiar with the firearm. His gun knowledge probably came from computer games rather than real-life shootouts, but that didn't reassure me. If he started blasting away like one of the soldiers in Battleground, we'd all be in big trouble.

"Put it back." I pointed at the drawer where the gun had been hidden. "We don't need it."

Ben laughed. "Really? You want me to turn the other cheek?" He snapped the revolver's cylinder back into place. "No, sorry. I'm not a saint. Just lock the RV's door, okay?"

Holding the gun in his right hand, he turned around and left the Winnebago. Emma obediently locked the door behind him.

This was a disaster in the making, but I couldn't stop it. Arguing with Ben would be pointless, especially now that he was armed. I propped my elbows on the dinette table and leaned my forehead against the palms of my hands. I was so, so tired. I could barely keep my eyes open.

Charlotte sat down beside me and put her arm around my shoulders. Then she looked up at Emma. "Listen, Joan needs some sleep. Is there any—"

"Oh my God, yes!" Emma pointed at the queen-size bed at the back of the RV. "She can sleep over there."

I shook my head. "No, no, I can't take your—"

"Don't worry, there's another bed right here, and it's just as big. I just have to fold down the dinette table and put a cushion over it." Emma helped Charlotte lift me to my feet. "I can sleep here with Marguerite and Charlotte. And that quiet boy…I'm sorry, what's his name?"

"Andrei," I muttered.

"Yes, Andrei can sleep in the overhead bed above the driver's seat. See, there's plenty of room for everybody."

I let Emma and Charlotte guide me to the bed, which was covered with a white sheet and a beige blanket. I was fading fast. They laid me down on the mattress and closed the curtain around it. That's all I remember.

* * *

I woke up at dawn, groggy and aching. Pale blue light seeped around the edges of the window shade and colored the bed. I knew that the morning session of the Science and Religion conference wouldn't start until nine, which meant I could sleep for another two hours, so I rolled over to a more comfortable position and adjusted my pillow. But before I could drift back to unconsciousness, I noticed that someone was in bed with me.

I was sleeping next to Charlotte.

I gave a start when I saw her there, and the bed shook. She was fully clothed, just like me, still in the yellow dress she'd been wearing ever since we left New York. She lay on her side, facing me, her right cheek mashed against her pillow, her eyes closed tight, her lips slightly parted. Her hands rested on the mattress, one on top of the other.

After a moment, though, her breath hitched and her eyelids quivered. I'd disturbed her slumber when I shook the bed, and now she was waking up too. She opened her eyes as I stared at her, and after another few seconds she smiled.

"Hey, Joan." She spoke in the softest whisper imaginable. "Hope you don't mind, but I switched beds a while ago." She yawned and stretched, straightening her legs and pointing her toes. "It was a tight squeeze on the other bed. With all three of us, I mean. And Marguerite kept sticking her elbow in my ribs."

I smiled back at her. I was glad she'd joined me, but it also made me nervous. For the first time since I met Charlotte, we were indisputably alone, with no one about to intrude. The curtain around the bed was closed, and judging from the silence in the RV, everyone else was fast asleep. I could say anything to her now, share my secret thoughts, confess my deepest feelings. But I was afraid. I didn't know where to start.

Just say what's on your mind! Tell her how you feel!

"I'm glad you're here," I whispered. "I wanted to thank you."

She lifted her head off the pillow. "Thank me? For what?"

I sat up on the bed. The mattress bounced a little. "You were the key, Charlotte. You got me out of New York, remember?" I raised my arm and pointed behind me. "And you got us the help we needed, the friends who helped us travel all the way here, thousands of miles. You made all of it possible."

She sat up too and wrapped her arms around her knees. "And I'd do it all over again. Because I believe in you. From the first second I saw you, I knew you were special."

This was a very nice thing for her to say, of course, but it wasn't what I really wanted. When Charlotte said "special," I got the feeling that she was thinking of revelations and prophecies and divine missions, and I didn't want her to think of those things right now. I wanted her to see *me*.

I reached for her. I rested my hand on her upper arm, just below the sleeve of her dress. "I want you to know something else. I haven't forgotten about your dad. As soon as we're finished with this conference, we're gonna drive to Santa Fe, to the New Life Oasis center. I already looked at the map, it's only an hour away." I slid my hand down to her forearm, which was downy and warm. "We're gonna get your dad out of that place. I promise."

Her eyes welled. She swallowed hard. "I…I knew you hadn't forgotten. You just had…you know, more important things to do first."

I shook my head. "No. Nothing's more important than you. I really like you, Charlotte. I mean, I really, *really* like you."

"And I like you, Joan. I feel the same way."

A tear slid from the corner of her eye. It glistened in the dawn light as it ran down her cheek, and it was beautiful, so beautiful. But I still wasn't exactly sure what she thought of me, not a hundred percent, not totally and completely. And because I'm a mathematician, I needed a rigorous proof. *I had to be sure.*

I grasped her hand and twined my fingers with hers. "The thing is, I like girls, okay? I'm a lesbian. And I'm hoping you like me that way too."

I squeezed her hand. For an agonizing three seconds, she said nothing. She just looked at me.

Then she squeezed my hand back.

"Come on. Isn't it obvious?" She raised her other hand and touched my face. "Haven't you noticed how much I've been flirting with you?"

She leaned closer and kissed me. Dizzy, exalted, I closed my eyes.

Chapter Twenty-Five

Two hours later I woke up a second time, because Andrei was shouting. His voice came from the front of the RV, on the other side of the curtain. To my great surprise, he sounded joyful.

"Joan! You won't believe it! They're all waiting for you!"

I opened my eyes. Charlotte's face was right in front of mine, our heads sharing the same pillow. She hadn't woken up yet, but her eyelids were quivering. One of her arms lay under the pillow, and the other curled around my waist.

Andrei's shoes stomped the floor of the RV. He came closer, heading for the back of the vehicle. "I just went to the convention center, and guess who I saw there? Edward Witten! The super-genius of string theory!" He stopped next to the curtain that ran around the bed. "And listen to this—he's looking for you! When I introduced myself to him, the first thing he said was, 'Are you a friend of the prodigy who calculated alpha?'"

I was confused. I'm never at my best when I wake up, and now too many things were hitting me at once. I lifted my head off the pillow, and Charlotte opened her eyes. At the same time, the curtain swung back and forth at the foot of the bed. Andrei was batting the fabric, trying to get my attention.

"Joan, are you listening? Witten saw the video of your speech at Cahokia, then checked with the researchers in California who'd measured alpha. Once he learned that you'd calculated the correct value, he tried to figure out who you were. He sent mass emails to physicists and mathematicians all over the country." He batted the curtain again. "And guess who responded to him? Professor Laura Taylor of City College! She'd seen the video too, and she told Witten that you were a serious mathematician. She said you had more than enough expertise to work on string theory. That surprised me, to tell you the truth. Remember how pessimistic she was during our meeting in her office? But I guess she changed her mind about you."

Now Charlotte lifted her head and turned toward the curtain. She looked just as confused as I was. It occurred to me that I ought to give Andrei a warning, maybe by saying something like "Hold on, I need some privacy, just give me a minute." But I hesitated too long, and I guess Andrei's impatience got the better of him. He reached for the edge of the curtain and yanked it aside.

"Joan, wake up! You need to…"

His joyful voice trailed off as he stared at us. Charlotte was in my arms, and I was in hers. We were fully clothed, but that didn't matter. Andrei only needed to look at our faces to see that we were a couple now.

"Uh, Joan?" He blinked a few times, as if trying to clear his vision. "Did you, uh, hear what I just said?"

I pulled away from Charlotte and sat up. Instead of answering him, I studied his face, trying to see how much damage had been done. Andrei wasn't frowning, but he seemed to be chewing his lower lip. It was like the expression he wore when he worked on a math problem, intent and inscrutable. Was he surprised? Maybe disappointed? Or was he shocked, crushed, furious? I couldn't tell.

Either way, though, I felt guilty. I should've been more honest with him. I should've had the courage to tell him who I was, right from the beginning.

"Andrei, I'm sorry. I—"

"There's no time to talk." He waved his hand, cutting me off. "Hundreds of people are waiting for you at the convention center. All the physicists and mathematicians, plus some of the philosophers and theologians too. I told Witten that you wanted to give your presentation as soon as possible, so he rearranged the schedule and added a special session for you at nine o'clock." He glanced at his watch. "Which means you have thirty minutes to get over there."

Andrei clearly didn't want to discuss the personal stuff, at least not right now. He was focused on our goal: unveiling the Theory of Everything, and maybe saving the world. And I realized that I should get focused too.

I nodded. "Okay. I just need a couple of minutes to brush my teeth." I turned to Charlotte. "You're coming with me, right?"

"Of course." She reached over and clasped my arm. "I'll sit in the front row and cheer you on. I'll start clapping every time you mention a new equation."

For the briefest moment, I thought I saw Andrei grimace. But then he quickly turned around and headed for the RV's door. "I have to make a few more arrangements. To keep you safe. For extra security, Witten said we could bring some of our supporters into the convention center. I'll meet you outside in five minutes."

* * *

Holding Marguerite's laptop and the two math notebooks under my arm, I stepped out of the Winnebago and into the center of a silent crowd.

Dozens of people stood on the asphalt of the parking lot, girls in peasant blouses and embroidered skirts, boys in ripped jeans and bandannas, all of them gathered around the RV. It was like a crowd at a protest march, except the protesters here didn't carry any "Stop the War" signs or chant any political slogans. They just stared at me as I stepped outside, blinking in the bright New Mexico sunshine. A few people smiled, but most didn't. Our business was serious, and the danger was real.

Charlotte, Marguerite, and Emma followed me out of the RV. To my left, Andrei huddled with a pair of young men in tie-dye headbands, giving them instructions in a low, anxious voice. To my right, Ben surveyed the crowd, looking in all directions, trying to inspect everyone there. He shaded his eyes with his left hand while keeping his right hand in the pocket of his jeans, where I glimpsed the outline of his revolver. The sight of it made my stomach clench. If an assassin showed up in Albuquerque today, I could at least take comfort in the knowledge that the Russian government taught its agents how to shoot straight. But I couldn't say the same for Ben.

After a few seconds, Andrei dismissed the two boys in headbands and came to my side. He faced the crowd and cupped his hands around his mouth. "Attention, everyone! We're ready to start! Now remember, we need to form a tight pack around Joan. If we shield her from view, no one can aim a gun at her. And as we march forward, don't let anyone infiltrate our group. Push the bystanders aside if you have to." He looked at his watch again, then raised his arm and pointed west. "All right, LET'S GO!"

The crowd set off. We moved slowly at first, in fits and starts, struggling to stay together as we crossed the parking lot. After the first minute, though, we fell into step, instinctively coordinating. We marched down Tijeras Avenue, dozens of people ahead of me and dozens behind, our bodies arranged in a V-shaped formation, with me at the very center. I heard shouts and cheers in the distance, but my escorts were shielding me so closely and effectively that I couldn't see any of the applauding bystanders.

Soon we made a right turn. Above the heads of the marchers I saw the tall windows of the Albuquerque Convention Center, which loomed over a broad, sun-dazzled plaza. The unseen cheering grew louder as we approached the center's glass doors, and a strange feeling came over me, a sharp jolt of dislocation. I felt like I was somewhere else entirely. I wasn't in Albuquerque, wasn't entering the convention center, wasn't even in America. I was in France, in the fifteenth century, stepping through the gates of a walled town. The people marching beside me were my knights and archers, my squires and swordsmen. My RESIST T-shirt was my armor, made of impenetrable steel, and the laptop in my hands was my shield and banner. I felt strong, confident.

We stormed into the building and climbed a wide stairway. Up ahead was another crowd, milling outside the doors to an auditorium. They were older and better dressed than the people marching with me. Some wore suits, some wore tailored dresses, and all of them had conventioneer badges hanging from their necks, displaying their names. They were the conference attendees, the researchers and intellectuals who'd been invited to discuss "Bridging the Gap Between Science and Religion." They looked impressive, even intimidating. But as we approached, they fell silent and stepped out of our way.

Andrei ran ahead and found someone to help us, a man wearing a radio headset and a black T-shirt that said "ABQ Audio/Video Specialist." The rest of us followed them into the auditorium and strode down the center aisle. Andrei led me onto the stage, where a lectern stood in front of a giant video screen. I handed Marguerite's laptop to the A.V. specialist, who set it on top of the lectern and plugged a cable into its USB port.

Meanwhile, my fellow marchers spread across the auditorium. Charlotte took a seat in the front row, and a dozen of our new friends sat behind her. Ben and several other marchers stood guard at the door, eyeing all the people who came inside, making sure that everyone had a conventioneer badge. At the same time, Marguerite and Emma walked down the aisles and examined everyone who was already seated. Our security plan wasn't foolproof, of course. A Russian agent could forge a badge, or steal someone else's, and slip into the audience unnoticed. But that was a risk we had to take.

The A.V. specialist finished hooking up the laptop to the video-screen controls. I stepped up to the lectern, opened the computer, and clicked on the PowerPoint file I'd created. The giant screen behind me displayed the first slide of my presentation, which showed the title—**THE THEORY OF EVERYTHING**—in large black letters against a white background. I wasn't an expert at PowerPoint, so my presentation had no fancy backgrounds or fonts. But I *did* create a nice computer graphic of my Calabi-Yau manifold, the same shape I'd drawn in my math notebook. I rested that notebook and Andrei's on the lectern, ready to use them as a backup in case the laptop malfunctioned.

Andrei leaned toward me. He cocked his head and grinned. "Everything okay, Joan? Are you ready?"

The look on his face was familiar. It was a big dopey grin, just like the one he'd flashed when I saw him for the first time, in Principal Barnes's office at my high school. Now that we were finally here, at our God-given destination, it looked like Andrei had let go of his fears. He no longer seemed so nervous about Vlad and his assassins and the fighting in Eastern Europe. He didn't seem upset about Charlotte either. We'd returned to the essential thing that had brought us together, our shared love for mathematics. We were about to reveal the most perfect theory imaginable.

I grinned too. "Oh, I'm more than ready. I'm *psyched*."

The A.V. guy double-checked the microphone on the lectern, tapping it. Then he went to the right wing of the stage and ducked behind the black curtain there. He reached for a switch on the wall and dimmed the auditorium's lights.

Andrei took that as his cue to leave. "Good luck," he whispered. "If you need me, I'll be right over there." Still grinning, he walked to the stage's left wing. Once he was behind the side curtain, he turned around so he could observe my presentation.

I glanced at my watch. According to the digital readout, it was 9:01:13, but several people were still filing into the auditorium, so I decided to wait another half-minute. I smiled at the audience, then lowered my gaze until I found Charlotte, sitting at the center of the front row. She saw me staring at her and waved.

Then I noticed who was seated to her left, recognizing him from one of the physics websites I'd visited. It was Edward Witten, Mr. String Theory himself, bespectacled and balding. And sitting to *his* left were two other string-theory pioneers I'd seen pictured on the website, John Henry Schwarz and Michael Green. Next in line was a tall, bearded man who also looked familiar, probably another string theorist—his name, according to his badge, was Matthew West. And I immediately recognized the woman at the end of the row: Lisa Randall, the greatest theoretical physicist of her generation.

It was overwhelming. I couldn't imagine a better audience. I closed my eyes and said a silent prayer of thanks. God had provided once again.

When I opened my eyes, I saw Ben closing the auditorium's doors. The last few latecomers were settling into their seats, so it was time to start. I leaned toward the microphone and pursed my lips, ready to say, "Hello."

Then I heard my father call my name.

"Joan! Hey, Joan! Over here!"

I turned automatically toward his voice, my head whipping to the left. Andrei had vanished from the left wing of the stage, and my dad stood in his place, behind the side curtain. He waved both hands in a frantic "come here" gesture.

My heart pounded. I pulled back from the microphone and almost lost my balance. I felt hot and lightheaded, shaking all over.

He saw the video of me at Cahokia. Dad heard where I was going and got on the first flight to Albuquerque.

"Please, Joan!" His voice was a fierce whisper. "We need to talk!"

Because Dad was in the stage wing behind the curtain, the audience couldn't see him, but they sensed that something was wrong. They shifted in their seats and stared at me, wondering why I just stood there, swaying behind the lectern. Charlotte gave me a worried look and half-rose from her seat. The string theorists glanced at each other and raised their eyebrows. Ben stepped away from the auditorium's door and came down the aisle, heading for the stage. I was making *everyone* nervous, but I couldn't help it. I couldn't think, couldn't move.

"Joanie!" Dad raised his hands to his head and clutched his hair. "*Please!*"

That snapped me out of my trance. I lurched toward the microphone and muttered, "Excuse me for a moment." Then I hurried across the stage.

As soon as I stepped behind the side curtain, Dad hugged me. He wrapped his arms around my back and kissed the top of my head. His relief was so strong, I could feel it in his arms, which trembled as he embraced me. And that feeling flooded me too, easing my nerves, steadying me. I thought of New York, our apartment on 78th Street. Dad smelled like home.

After a few seconds he pulled back but still gripped my shoulders, holding me at arm's length. "Oh Joanie, you had us so worried! Mom and I had no idea where you went!"

I couldn't speak. I just looked at him. His hair was a mess because he'd tugged at it so wildly. And had it gotten a little grayer since the last time I saw him, just three days ago?

Yes. It looked grayer.

"But it's all right now, Joan. Everything's all right. We just have to go downstairs and tell Mom you're okay. She's across the street, at the hotel." He slipped an arm around my waist. "Come on, we can get out of the building this way. There's an emergency exit."

I let him lead me a couple of yards farther from the stage. Then I stopped. "Uh, Dad? I…I have to go back to the, uh…" I pointed over my shoulder. "Back to the lectern. They're waiting for me."

He furrowed his brow. "Waiting for you?"

"Well, you saw the audience, right? I'm about to give them a presentation." I turned around and pointed at the video screen. "About the Theory of Everything. I solved the problem, Dad. That's why I'm here."

I found it strange that I needed to explain this to him. If Dad had watched the video of my speech at Cahokia, he should've known why I'd come to Albuquerque. But he just looked at me blankly. After a few more seconds, he frowned.

"Joan, your mom's been crying nonstop ever since you ran off. After you left the psychiatrist's office, we heard gunshots outside and a car skidding down the street. When we called the police, they said it was possible that you were kidnapped. How do you think that made us feel?"

"Look, I'm so, so sorry about that. But right now I need to—"

"No, that's enough. We're going straight to your mother." He tightened his grip around my waist. "She has to see you. I won't let her wait a second longer."

He tried to pull me toward the emergency exit, but I dug in my heels and leaned backward. "Dad, stop! You're not listening to me! I need to give this presentation, okay? As soon as I'm done, I'll go right to the hotel and see Mom. Or you can bring her over here and—"

"Enough, Joanie! You've put us through enough!" He let go of my waist and grabbed both my arms. Then he yanked me forward. "Come on!"

I lost my footing and dropped to my knees. "Stop! *Stop!*"

But he kept pulling me, dragging me across the floor, his fingers digging into my forearms. He'd never done anything like this before, *never*. I felt nauseous, appalled, sick to my stomach.

He's hurting me! How could he do this?

Then it hit me. I realized what was going on. It was so simple.

This wasn't my father.

And when I looked up at him an instant later, it wasn't Dad gazing down at me. He'd taken a different shape, a tall, muscular body in a white robe, a craggy face with a long gray beard. It was the face I'd seen in so many churches, the image that had been painted on the ceiling of the Sistine Chapel and a million other places. It was the wizened Old Testament face that everyone expects to see in Heaven, the ancient, glowering face of God. His eyes blazed like sapphires.

"I apologize for the deception, Joan. I was hoping to stop this nonsense with a minimum of fuss, but I see now that you're too stubborn and clever. You're very determined to reveal this theory, aren't you?"

I gaped at Him. I'd never been so confused in my whole life. "Isn't that what you want? The revelation?" I pointed at His gray beard, His blazing eyes. "Isn't that why you told me to come here? So I could unveil your design?"

He shook His head. "No, that's not what I want. In fact, that would be a catastrophe. And the truth is, I've never spoken to you until this very minute."

"What are you talking about? You spoke to me *four* times!"

"No, I'm afraid you're mistaken. On all those occasions, you were speaking and making plans with someone else."

I shivered on the floor. "Someone else?"

"Yes, the Adversary. The Prince of Lies."

Then He laid His hand on my forehead, and I lost consciousness.

Chapter Twenty-Six

"Wake up, Joan."

No. I didn't want to. I wanted to stay in the darkness.

"You can't hide from this. You've caused a great deal of trouble."

I did nothing wrong. Nothing immoral or irrational.

"I'm not accusing you of any crime. You were misled. But deliberate or not, your actions have consequences."

I wanted to sleep. I wanted to forget the world and think only of pure mathematics, numbers and equations, graphs and functions.

"*Wake up!*"

I couldn't resist any longer. I opened my eyes.

I lay at the foot of a gigantic sand dune. Its steep slope loomed above me, so tall that it cast a bluish shadow over the patch of ground where I'd materialized. For a couple of seconds I just stared at the fine white sand under my body. The grains stuck to my jeans and T-shirt like powder. Then I sat up and looked around.

Hundreds of huge sand dunes lapped the landscape. They stretched like solid waves in a vast white ocean, their slopes shining under the brilliant sunlight, their ridges outlined against the blue horizon. The dunes rippled and twisted, frozen in place and yet eternally moving. It was beautiful and silent and otherworldly. I'd never seen anything like it, so clean and white.

And standing a few yards in front of me, just beyond the edge of the dune's shadow, was the artist who'd sculpted this landscape, the Creator Himself. His robe draped His brawny eight-foot-tall body, the white fabric hanging straight down in the still desert air. Dozens of sand grains clung to His hair and dotted His long gray beard. Squinting, clearly irritated, He looked down at me.

"Are you fully awake? And listening? I need your undivided attention."

I took a deep breath, then rose to my feet. I felt weak and lightheaded. The desert seemed to tilt for a moment, the sand dunes yawing. I blinked several times to steady myself.

"Where are we? Is this Heaven?"

He shook His head. "No, you're on Earth. In New Mexico, to be precise, but a hundred and fifty miles south of Albuquerque. This is the Tularosa Basin, the White Sands desert."

That sounded vaguely familiar. "Hold on, I've heard that name before. Where—"

"Your friend Charlotte mentioned it. This is where her father used to hold outdoor services for his homegrown church."

Now it came back to me. Charlotte had said it was a spiritual place, a good spot for getting in touch with the "cosmic consciousness." She was more right than she knew. "Why did you bring me here? What's going on?"

"We needed a quiet place to talk. I've taken steps to guarantee that no one can interrupt us." He lifted His right arm and pointed straight up. "Take a look."

I raised my eyes. Near the sky's zenith I glimpsed a jet airliner, at least five miles up, appearing tiny against the blue expanse. Trailing behind it was a long white streak, the contrail of ice crystals that form in the wake of the jet engines. The streak was sharp and slender just behind the plane and much wider and fuzzier farther back.

But the jet didn't move. I stared and stared at it, but the plane didn't budge.

Alarmed, I took a step forward, then another. Then I emerged from the dune's shadow and turned around to look at the sun.

A warped halo blazed around it. Blinding beams twisted across the eastern half of the sky, diving toward the horizon and curving around the morning sun, making dozens of convoluted loops. It was the same halo I'd seen around the full moon after the homeless woman stopped time at 125th Street. And now time had stopped again. I felt it in the silence of the desert, the stillness of the air.

The Lord Almighty lowered His arm. "No time has passed since I removed you from the convention center. All the scientists are frozen in their seats in the auditorium, and the Earth has stopped spinning. I've paused the entire cosmos so we can have this conversation, Joan. That's how important it is."

I tried to give Him my full attention, but it was impossible. I couldn't help but stare at the golden halo, the shape that matched the Calabi-Yau manifold. I pointed a trembling finger at it. "She did the same thing. The homeless woman on the subway. She made everything freeze."

"That was the Adversary. He also disguised himself as the police captain and the runner in the park and the goateed man in the tattered jacket. He's powerful, and he knows the secrets of the universe as well as I do. But his intentions are different from mine. His goal is chaos."

"So he's the devil? *Satan?*" My throat tightened. "I was talking with Satan all that time, and I didn't even know it?"

The Almighty nodded. His robe swayed around His legs. "He has many names and disguises. It's not your fault that he tricked you. He deceived me too, hiding all his machinations from my sight. I didn't discover his plot until a few minutes ago, when you entered the convention center. I rushed in to stop you when I saw that you were about to reveal the theory."

I felt lightheaded again. I was so confused. "This whole thing was a *plot*? Cooked up by the devil?"

"It isn't the first time. The Adversary delights in schemes that cause dissension and disorder. He recruits human beings as his agents and uses them to stir up trouble." God frowned in distaste, twisting His thick lips. "He has a long history of manipulating people, tempting them with their fondest desires. In your case, he preyed on your human weakness."

"I'm sorry, I don't—"

"You were distressed. The death of your sister left you angry and disoriented. In your bewilderment, you thought you could use your mathematical abilities to regain some control and lift your battered spirits." He looked straight at me, narrowing His sapphire eyes. "You wanted to prove your worth by solving the most difficult problem you could find. And that's how the Adversary tempted you. You obeyed him because he offered you the Theory of Everything."

He took a step toward me. I didn't like what He was saying, but I accepted it. He was God, so He knew every corner of my mind, knew me better than I knew myself. Maybe I'd hidden my true motivations. Maybe I'd fallen for the devil's lies because I'd wanted to believe them so badly.

But after a few more seconds, I shook my head. The Almighty's explanation didn't sound right. "No, it was more than that. He told me that billions of lives were at stake. And then the war started in Europe, just like he said it would, and I thought I was doing the right thing by—"

"More lies. Don't you see how clever he is? He analyzes my plans so he can take advantage of them." God raised His voice, and it echoed against the dunes. "Let me tell you something about the Adversary. He knows enough about this world that he can guess its future, at least in a limited way. He can predict certain important events a few days in advance, mostly because he's so familiar with the diseased minds of the world's leaders. So when he spoke with you in New York he knew that a small war was imminent. But he also knew that this war would stay small. It wouldn't set off a bigger conflict or lead to global destruction. When he warned you about billions of deaths, it was a malicious exaggeration."

"So we're *not* on the brink of World War Three?"

"The conflict is already ending. Ten minutes ago, just before you arrived at the convention center, the American president sent a message to the Russian leaders, a proposal for a compromise."

"What? I thought—"

"The White House had no choice. The Russians were too fast and cunning, and there was no feasible way to force them to retreat. The compromise will allow them to occupy the Baltic States, and in return Russia will promise not to invade any other countries. Tempers will cool and the ballistic-missile alerts will be called off. Within a week, the crisis will be over. Within a year, the public will largely forget it."

I was glad to hear this news, of course. For the past two days I'd been terrified of nuclear war, panicking at the thought that I'd be unable to stop it, so it was a great relief to learn that there was no looming threat of an apocalypse. But I was still confused.

"Why did the Adversary exaggerate the danger? Did he want me to be scared?"

The Lord Almighty nodded. "As I said, he takes advantage of weakness. He knows that human beings are driven by fear. He frightened you to ensure that you would follow his orders." He frowned again and took another step toward me. "If you hadn't been so afraid, you might've seen through his deception. You might've scrutinized his absurd claim that the revelation of the Theory of Everything would usher in a golden age of peace and harmony. But your fear blinded you."

The Almighty stood less than a yard away, looking grim. He gazed down at me as if I were an ignorant child, a failing student, an irredeemable creature. I saw no divine love in His brilliant blue eyes, only disappointment. At first I felt ashamed and tried to stammer an apology, but I couldn't get the words out. I was struck dumb by the unfairness of the situation. *Why should I apologize? What exactly did I do wrong?*

I raised my head and returned His stare, daring to lock eyes with Him. "I don't think the plan was absurd. I was going to reveal the Theory of Everything to the scientists and spiritual leaders at the conference. They're responsible intellectuals who could understand the theory and spread the news to the whole world." I pointed north, toward Albuquerque, where my audience sat frozen inside the convention center. "For the first time in history, people would have *proof* of the universe's divine origin. And who knows, maybe this revelation would change the world for the better. That doesn't seem so improbable."

My stomach churned. I was questioning the wisdom of God, and He clearly didn't like it. He furrowed His craggy forehead, moving His gray eyebrows closer to each other. "Were you there when I created the universe, Joan? When I unleashed the Big Bang and ignited the first stars?"

"No, but I—"

"Then don't talk to me about probabilities. I've observed the human species ever since it evolved from the apes, so I know its character far better than you do. Humans will turn *anything* into a weapon. If you gave them the divine blueprints of the universe, what's the first thing they'd do? They'd use the plans to figure out a way to demolish the cosmos."

"Okay, that might be true of some people, but not everyone. That's why I went to the conference. I was going to share the theory with peacemakers, not warmongers. I planned to—"

"Enough!" The Almighty sliced His arm through the air, and His fingers came within inches of my nose. "The Adversary has deceived you so thoroughly that his lies are spewing out of your mouth. You need to forget everything he told you, every last word. If you don't, you'll become just as perverse as he is."

God had grown furious. His wrath was fearsome to behold, a thunderstorm raging behind His face. I stepped backward, retreating a couple of yards, but I didn't take my eyes off Him, not even for an instant. I was angry too, and I wanted Him to see it.

"What about *your* lies? Back at the convention center, you disguised yourself as my father. You tried to deceive me just like the Adversary did."

He scowled. "No, my goal was very different. I was trying to restore some order, to make things right. And that's what I'm still trying to do now, even though you're stubbornly resisting it. I have to repair the damage that's been done." He pointed at me. "Do you realize that the Adversary poisoned you? That he tricked you into swallowing a substance that made you more susceptible to his lies?"

I took another step backward. "Poisoned?"

"Remember the tea you drank four nights ago? Just before you drew the Calabi-Yau manifold on the window of your sister's bedroom?"

I folded my arms across my chest, instinctively trying to protect myself. I remembered the orange teabags and the taste of the tea, like a salty apple. "What? How did—"

"The Adversary wanted to keep a constant watch over you. Speaking with you on the subway train and in the police car wasn't enough. So he took another human form, the shape of a young man, and he became your friend. He encouraged you to work on string theory. And he gave you the poisoned tea, which altered your mind so you could recognize the mathematical solution he'd shown you. He pushed you down the path he wanted you to take."

My heart started pounding. My chest was so tight, I could hardly breathe.

The Lord turned to the left and raised His arm. He pointed at the top of a nearby sand dune, about a hundred feet away. "I wasn't sure if you'd believe me, so I ordered the Adversary to meet us here. I'm sure he'll be happy to answer all your questions."

A moment later, a pale blond boy appeared at the top of the dune, dressed in rumpled khakis and a polo shirt. Andrei Mishkin looked right at me and bowed.

Chapter Twenty-Seven

With a loud whoop, Andrei leapt off the dune's crest. He bounded down the steep slope, his ugly brown shoes splashing the sand and leaving a trail of shallow craters in the whiteness. Laughing, he scrambled and skidded to the wide trough that ran between the dunes. Then he grinned and dashed toward us. He stopped right in front of me, so close that his last steps kicked sand all over my sneakers.

"Joan! I see you've met the Big Cheese!" Andrei jerked his head toward the Almighty. "He doesn't like it when I call Him by that name, but it's appropriate, don't you think?"

I stared at him in disbelief. He was treating God like a fraternity brother, but that wasn't nearly the worst of his sins. He'd deceived me! He'd pretended to be my friend and handled me like a puppet! And worst of all, he'd *poisoned* me! How in the world could he even talk to me now, much less grin that dopey grin of his? Where did he get *the nerve?*

I opened my mouth, ready to curse him out, but I couldn't say a word. I was aghast.

Andrei seemed completely oblivious to my feelings. He laughed again, then turned to the Lord. "Greetings, oh King of Kings! Oh Rock of Ages! You have to admit, I came pretty close to succeeding this time, didn't I?"

God didn't say anything either. The expression on His face was a mix of disgust and resignation. Although He clearly detested the Adversary, the hatred was so longstanding that it must've lost its edge. What was the point of denouncing His enemy yet again? Better to save His breath.

But Andrei appeared to be eager to talk. He shifted his weight from foot to foot, joyful and antic. "Yes, I bet you were worried once you saw what I was up to. Sharing the knowledge of your cosmic design with human beings? Giving them secrets that would let them tap the hidden energy of the universe? It must've frightened you, eh? It probably scared you enough to curl your beard, am I right?"

The Lord grimaced. He averted His eyes from Andrei and looked at me instead. "Now you can see why he's called the Prince of Darkness. His greatest pleasure is tearing down what I've built."

Andrei looked at me too. He was still grinning. "The Prince of Darkness! To tell you the truth, I like that name. It makes me sound so intimidating. Like a creature in a horror movie!" He made a scary face, widening his eyes and baring his teeth. Then he laughed. "But no, it's not accurate. In reality,

I'm more like the God of Impulsiveness. Or better still, the God of Creativity. I like to shake things up a little."

The Almighty shook His head. "He's the Destroyer. The Chaos Bringer."

"Yes, I admit it, sometimes I destroy things." Andrei shrugged. "And yes, I sometimes mess up the Big Cheese's plans. But it's always for a good cause, you know? Believe me, if I stopped doing it, the world would be a much less exciting place. This whole universe would become as boring as Facebook."

He laughed again and nudged me with his elbow, obviously hoping that I would laugh too. But I was as far from amused as I could possibly be. Andrei didn't seem godlike at all. He reminded me of the annoying boys at my high school, the kids who loved to play pranks and torment anyone who was even slightly different from normal. If you called them on it, they'd always say, "It's just a joke! Can't you take a joke?"

I glared at him. "You're evil. You don't care what happens to anyone, do you?"

The Lord nodded in agreement. "He doesn't. He roped you into his scheme by warning that billions of people would die if you didn't cooperate, but that was the very opposite of the truth. Because the Theory of Everything would show scientists how to build new weapons, its revelation would inevitably lead to devastating wars. But that prospect didn't bother the Adversary. He thought it would just make the world more 'exciting.'"

Andrei finally stopped grinning. He raised an eyebrow and stared at the Almighty. "First of all, the devastation is *not* inevitable. You know very well that a more positive outcome is possible too, especially if the revelation is handled correctly. And second, how many people will die as a result of *your* schemes, oh Pillar of Fire?"

God gave him a dismissive look. "I don't scheme. I have a carefully constructed plan. There's a difference."

"Well, let's leave the semantic argument for another time, shall we? My point is that your careful plan for humanity's development will also take a horrible toll on the species. Under your guidance, how many billions of people will be slaughtered before the end of the 21st century? Massacred by climate change, pandemics, starvation, genocide? And by the weapons that human governments already possess?"

The Lord grew wrathful again. His anger stirred the desert air, making it swirl and crackle. His robe flapped in the wind He'd made, and the sand grains whirled around His bare feet. He stepped closer to Andrei, towering over him. "We're not the same. You trigger catastrophes because you find them entertaining. They have no other justification. But I have an aim, an objective."

He raised His right hand and the wind blew harder. A powerful gust whistled past us and struck one of the nearby dunes, which exploded into a thick cloud of sand. The grains flew upward and hung in the air for a long moment, blocking the sun and its warped halo. Then the sand settled down in an orderly way, piling on the ground in a distinctive geometric shape.

The Almighty pointed at the descending streams of sand. "All my actions are intended to make the universe more majestic. I've pursued this goal from the beginning, starting with nothing but a simple burst of energy. And now, after fourteen billion years, look at all the wonderful complexity I've created." The last of the sand grains fell into place. The shape on the ground was a terraced pyramid with hundreds of perfectly formed layers. "The process is long and difficult, with many setbacks and painful choices along the way. But the endpoint will justify all the sacrifices."

Andrei was clearly unimpressed. He glanced at the pyramid of sand for half a second, then turned back to me. "I suppose it's reassuring that the Creator is so perseverant. But His goal isn't something that the average person would want. He's not interested in maximizing human happiness. He's a God of order, not a God of love."

"That's another lie!" God thundered. "The human beings who survive this century will eventually form a more sustainable civilization. They're an important part of my plan."

"But that's not much consolation for the billions who *won't* survive." Andrei kept his eyes on me. The look on his face was sympathetic, almost pitying. "And I hate to give you more bad news, Joan, but there's no afterlife for humans. As you can see, the universe is complicated enough with just two immortal beings."

The Lord shook His fist. "And how I wish there was only one!" Enraged, He stamped His bare foot on the ground, and the impact shook the desert. I lost my balance and fell to my knees, and the beautiful pyramid collapsed. Within seconds, it was a sand dune again.

Andrei bent over me and offered his hand. "See what I have to deal with? Luckily, the Big Cheese can't get rid of me. He's stronger than I am, but for some reason he doesn't have the power to finish me off. And maybe that says something interesting about the universe, eh? Maybe you can't have order without chaos, and vice-versa. What do you think?"

I didn't say anything at first. I didn't take Andrei's hand either. I just stared at it. I was so baffled, I needed to focus on something solid, something ordinary. It was the hand of a math geek, its index and middle fingers bluish with pen marks, its nails bitten to the quick.

Finally, I looked up at him. "I don't understand. God is supposed to be good. He's supposed to care about people, to want what's best for us. And the devil is supposed to be evil."

Andrei stood up straight and sighed. "Yes, that's what your religions say. But that assumption doesn't make a lot of sense when you think about it. The concepts of good and evil are applicable only to intelligent living things—a rock or a plant can't be evil, correct? And intelligent beings are incredibly rare in this universe. That's why your scientists have failed to detect any extraterrestrials. Even on Earth, good and evil had no meaning until humans evolved, which was a very short time ago compared with the age of the planet. If God were truly interested in goodness, He would've built an entirely different kind of cosmos."

I turned away from him and looked up at the Almighty, expecting an angry rebuttal. But He merely shrugged. "I created you, didn't I? Wasn't that good enough?"

Andrei pointed at Him. "He cares about you in the same way that a scientist cares about his lab rats. Your well-being is less important to Him than the results of His experiments." He pulled his arm back and pointed at himself. "And I've been misrepresented in your religions too. I'm not an evil spirit who wishes you harm. I just like to argue and play games and dress up. I've thoroughly enjoyed being Andrei Mishkin for the past two months."

I rose to my feet and looked him in the eye. His last words didn't make any sense. "Two months? I saw you for the first time just a week ago."

He grinned again, his clownish cheerfulness returning. "Well, I had to lay the groundwork for the masquerade, yes? I found my opportunity in the Russian city of St. Petersburg. A very troubled computer scientist jumped off a bridge into the Neva River, and the man's son dove into the water in a futile attempt to save him. The son drowned too, but before his body floated to the surface I put my spirit into the corpse and reanimated it. Then I swam to shore and rushed to the boy's home. I made the switch so quickly and neatly that the Big Cheese didn't notice what I'd done."

The Almighty winced. "What a loathsome trick. Your depravity never ceases to amaze me."

"Oh, you're just too squeamish. It wasn't so terrible. In fact, I was a great comfort to the boy's mother as she grieved over her husband's death. And then I arranged their immigration to America by making a deal with that ugly Russian spy, the one who'd worked with the boy's father."

I bit my lip. I was starting to see the depth of the Adversary's deception. "So Vlad is a real person? He's not one of your disguises?"

"Colonel Vladimir Karamazov is a real human, one of the more repugnant members of your species. But despite his bad manners, he was useful. He motivated you to embrace your mission, because you had an enemy to fight. I've observed that human beings rarely rise to their full potential unless they're engaged in conflict."

Andrei's voice was light and jaunty. Although his scheme had failed in the end, he seemed to really enjoy talking about it. But I wasn't enjoying his recap, not one bit. "So this was just a game for you? Pretending to be God? And then making me think I was crazy for believing it?"

He bowed to me again, very formally this time. "I offer my sincere apologies, Joan. You have to admit, though, the experience wasn't all bad. You've also had some pleasant moments over the past few days, eh?" He leaned closer and winked. "You and Charlotte seemed very happy together this morning."

I wanted to slap him. Rage burned inside me. A tide of blood surged up to my skull and washed the backs of my eyes. I literally saw red. "Now I know why God hates you so much. You're…you're…"

"Okay, calm down. I'm going to set things right." Andrei stepped away from me and addressed the Almighty. "Oh Fountain of Living Waters, hear my plea. Gaze upon Joan Cooper, the poor deceived creature who thought she was serving you. Don't punish her for the transgressions I urged her to commit. Take pity on Joan and return her to her family."

The Lord said nothing. He folded His arms across His chest and looked down at the Adversary.

Andrei squared his shoulders and tried again. "We can make an arrangement that will satisfy your concerns about the Theory of Everything. I'm sure you have the power to guarantee that Joan won't ever reveal the theory to another human being. That should be a simple task for a deity as formidable as you, correct?"

The Almighty kept him waiting for another five seconds. Then He shook His head. "No, it's not that simple."

Andrei furrowed his brow. He looked askance at the Lord. "What's the problem? Couldn't you just erase the theory from her mind? By readjusting the brain cells that hold her memories?"

My breath caught in my throat. I couldn't believe it. *They want to mess with my brain?* But before I could express my shock and anger, the Almighty shook His head again. "This isn't an ordinary memory. When Joan found the solution to string theory, it was like learning how to walk or speak or play music. She acquired a completely new skill, and that kind of information is deeply embedded in the human brain." He unfolded His arms and pointed at my forehead. "If I erased that skill from her cells, it would maim her personality.

She'd lose her mathematical aptitude and her zest for discovery. In short, she would no longer be Joan Cooper. She'd become a pale shadow of herself."

I'd heard enough. I wasn't going to let them lobotomize me. "Look, you don't have to do anything to my brain. I promise I won't share the theory with anyone. I'll burn the math notebooks and delete the PowerPoint file from the laptop. I'll take the secret to my grave, okay?"

The Lord rolled His eyes. He clearly thought I'd just said something ridiculous, and He didn't even bother to hide His disdain. "You expect me to rely on a promise from a human being? Do you have any idea how many promises your species has broken?"

Andrei stepped forward, moving between me and the Almighty. "Yes, the human race is generally unreliable, but Joan is an exception. I've spent the past week in her company, and I can assure you that I've never encountered such an honest, trustworthy person."

"Oh really?" God frowned. "Since when did the Prince of Lies become an expert on honesty?"

"Please, just listen to the facts. When Joan Cooper commits to a task, she always follows through. When I told her to go to the conference in Albuquerque, she overcame a hundred difficulties to get there. She made a plan and sought help from her friends and pushed aside every obstacle. I was especially impressed by how she handled the state trooper in Tucumcari. Even the devil—I mean myself—couldn't have swayed that upstanding man. Yet Joan managed to do it."

Andrei turned sideways so he could simultaneously look at the Almighty and point at me. He'd dropped his jaunty tone and now spoke in a serious, impassioned voice, arguing on my behalf. He was acting like my lawyer, and I guess that was appropriate for the occasion. I was definitely on trial. I'd allowed myself to be duped and manipulated, and now I was in serious trouble. My fate rested in the hands of the Judge of Judges, who had a long history of wrath and retribution.

I shuddered. *Joan of Arc was also put on trial, after the English captured her. And it wasn't a fair trial. Her enemies were determined to kill her.*

My own trial wasn't going much better. The Lord stared at the Adversary and narrowed His eyes, clearly unpersuaded. "You're leaving out some of the facts. In the past few minutes I've examined the details of your scheme, all the devious maneuvers you concealed from me, and I see that Joan wasn't exactly a model of truthfulness. She tried to hide her sexual orientation from you. And she deceived her parents when she left New York, using her psychiatrist appointment as a cover for her escape. If she could break faith with her mother and father, what's to stop her from doing the same to me?"

I felt a twinge in my gut. He was right: I'd treated my parents horribly. By now they'd probably seen the video of me at Cahokia—it had been on the Web for the past 24 hours—but it was also possible that they were still in the dark, waiting in their apartment for a call from the police, hoping for any kind of news whatsoever. And what made it so much worse was that they'd already gone through hell when Samantha died. They were traumatized and psychologically fragile. If they lost me too, they might never recover.

I couldn't let that happen to them. More for their sake than for mine, I humbly approached the Lord. I halted at a respectful distance from Him and lowered my gaze. Then I knelt on the sand in front of Him.

"Please, have mercy. I won't break my promise. Let me go home."

I waited on my knees, in silence, looking down at His bare feet. I was pleading for my life, and I hated doing it, but I had no choice. This was a patriarchy I couldn't smash, a divine power I couldn't defeat.

The silence stretched. The sand dunes sighed. The sun and its warped halo blazed above us.

Then the Lord passed judgment. "I'm sorry, Joan. I can't take the risk. My plans will be ruined if you share what you know."

I kept looking at the ground. I was too stunned to rise. It was a death sentence.

After a moment I heard footsteps to my right. Then I saw Andrei's brown shoes angrily stomping the sand. He stopped beside me and faced the Almighty. "What do you intend to do with her, oh Author of Peace? Stone her? Impale her? Or will it be the crucifix again?"

The Adversary's voice had deepened to a low, sarcastic snarl. He didn't sound like a seventeen-year-old Russian boy anymore. He was so full of outrage, he didn't even sound human. I looked up and saw him glowering at God, his face red, his eyes shining.

But the Lord stood His ground. He clenched His hands, and His robed body grew larger, expanding right in front of my eyes. His torso swelled to the size of a rooftop water tank, and His legs grew as long and thick as telephone poles. In seconds he was more than twenty feet tall, and his head was level with the tops of the dunes. God was far more powerful than the Adversary, and both of them knew it. It looked like He was ready to pound Andrei into the sand.

He definitely wasn't a loving God, at least not now. He was monstrous. His voice boomed over the dunes.

"I'll do nothing to harm her. I'll simply return her to the stage of the auditorium in Albuquerque." He pointed a giant finger at Andrei. "The harm will come from *you*. From the cruel forces you set in motion when you launched your unholy scheme."

I stood up, rising slowly and carefully. The ground rumbled under my feet, but I kept my balance. "What do you mean?" I tilted my head back and shouted to get His attention. "What forces?"

"Do you remember the convention center's audio/video specialist? The man in the black shirt who set up your laptop at the stage's lectern? You saw him just before I appeared in the auditorium, so he should be fresh in your memory."

I remembered the man but had no idea where this was going. I exchanged looks with Andrei, but he seemed just as confused as I was.

The Almighty shook His huge head. "That man was only pretending to be an employee of the convention center. He's actually a Russian agent, one of Colonel Vladimir Karamazov's men. He was assigned to kill you, so he planted a hidden bomb in the lectern."

All at once, I felt cold. The desert darkened. I was losing consciousness again.

"Wait!" I couldn't see God anymore. I couldn't see Andrei either. "*Stop!*"

"I'm sending you back to the auditorium now, and the flow of time will resume. Exactly ninety seconds after you get there, the bomb will explode."

Chapter Twenty-Eight

It was a kind of divine teleportation, I guess. In an instant I was back at the Albuquerque Convention Center, standing in the exact same spot where I'd left it. Because this was the second time God had instantaneously transported me, the experience didn't surprise me as badly as it had before, so I was relatively clearheaded when I reappeared on the left wing of the auditorium's stage. And that was lucky, because I had an important choice to make.

I looked at my watch. The digital readout said 9:04:28, and the clock was ticking. I had more than enough time to flee the auditorium before the bomb went off, but the Almighty had known that I wouldn't choose to do that. How could I? Charlotte sat in the front row, less than ten feet from the lectern. Sitting next to her were some of the world's greatest scientists and mathematicians, not to mention all the people who'd come here to support me, Marguerite and Emma and dozens more. I couldn't run out on them. I had to get them to safety.

Without waiting another second, I charged across the stage to the lectern. The audience gaped at me when I emerged from behind the side curtain, everyone wondering what was going on and why I was running. I glanced at the other side of the stage, where I'd last seen the A.V. specialist who was really a Russian agent, but he was long gone, probably out of the building by now, and when I reached the lectern I gave it a quick once-over, but there was no time to search for the explosives he'd hidden there. Instead, I grabbed the microphone and tried to keep my voice steady.

"Everyone, I've just been informed of a bomb threat. Please walk *calmly* out of the auditorium. There's no need to panic, just head for the nearest exit."

A burst of shocked cries rose from the audience. Some of them immediately stood up, while others craned their necks and twisted around in their seats, suddenly alarmed and trying to pinpoint the danger. The quicker-thinking members of the crowd followed my instructions and made their way to the aisles, but the less attentive people just stood there, paralyzed, in front of their seats. They blocked the auditorium's curved rows, preventing everyone else from leaving.

I needed help. I searched for familiar faces in the crowd and found them: Charlotte stood right in front of the stage, anxiously looking up at me, while Marguerite and Emma stood behind the last row, and Ben marched down the center aisle. One by one, I pointed at them and shouted orders into the microphone.

"Charlotte! Take all the people in the front row to that exit by the stage! Ben, get them moving down the aisle! Marguerite, open those doors at the back!"

They snapped to it, fast and efficient. Charlotte got the attention of the confused people in her row, including Edward Witten and the other string theorists, and led them to the front exit. Marguerite and Emma opened the back exit and waved dozens of people through the doors. Ben nudged everyone along and shouted at the laggards blocking the aisles. There were a couple of small accidents—Matthew West, the bearded scientist in the gray suit, pushed his way past Witten, who dropped his glasses—but most of the crowd filed into orderly lines and moved quickly out of the auditorium.

I checked my watch again—9:05:19—and jumped off the stage. To my great relief, the room was emptying. Only a dozen people were left, all of them within a few yards of the back exit. I ran up the center aisle toward them, intending to shove the stragglers through the doorway and follow them out of the room. Then we'd all be safe. The bomb would go off in forty seconds, but it wouldn't harm anyone.

I sprinted past the rows of empty seats, breathless and triumphant. I'd outsmarted the Russians. And God too.

Then I heard a loud, distinctive noise behind me. A gunshot.

Something hit the back of my leg, halfway up my right thigh. I tumbled forward, looking down as I fell. The bullet had ripped through my leg. Blood sprayed from a hole in my jeans, just above my knee.

I hit the floor hard, breaking the fall with my forearms. I didn't feel any pain in my leg yet, but I knew it was coming. In just a second, maybe two, it would annihilate me. Panicked, I flipped onto my back and raised my head from the floor of the aisle, scanning the front of the auditorium to see who'd fired the shot.

It was the scientist with the beard and the "Matthew West" convention badge. He came toward me, holding a black pistol in his right hand as he marched up the aisle. At the same time, he raised his left hand and pinched the edge of his beard. In one swift motion, he pulled a mask off the lower half of his face. It was an adhesive cloth covered with thick, curly beard hair.

As soon as he was beardless I realized why'd he'd looked familiar. My earlier assumption—that I'd seen his picture on a string-theory website—was wrong. The man wasn't really a scientist. And his name wasn't Matthew West.

It was Vlad.

The pain hit me as soon as I recognized him. I stared at the Russian and screamed.

Vlad stopped beside my writhing body. He checked his watch, then looked down at me. "Thirty seconds left. How did you find out about the bomb, Little One?"

I couldn't have answered even if I'd wanted to. All I could do was scream. The pain was unbelievable.

Vlad shrugged. "Well, it doesn't matter. This is my backup plan. Shoot you and run." Grinning, he raised his pistol. "I warned you this would happen. But you thought you were so clever."

He pointed the gun at my head. I kept screaming.

Then a second gunshot echoed against the auditorium's walls.

But I felt nothing this time. Not even the slightest jolt.

Vlad winced, looking surprised and confused. He stuck his left hand inside his suit jacket, as if reaching for his wallet. When he pulled his hand out, it was slick with blood.

He stared at it for a moment, uncomprehending. Then he crumpled to the floor.

Ben stood by the stage, twenty feet away, holding a revolver with both hands. It was the gun that Emma had given him the night before.

"Joan!" He lowered the revolver and ran toward me. "I got him! He was a Russkie, right?"

The pain was so tremendous, I couldn't think straight. Blood gushed from my leg, out of the entrance wound in the back of my thigh and the exit wound in the front. My jeans were soaked with it, and I couldn't stop shivering, but for Ben's sake I struggled to rise above the pain. I had to get the words out. I had to warn him.

How many seconds are left?

"Ben…you need to…"

"I know." He bent over me and slipped his arms under my legs and back. "I'm gonna get you out of here. Just—"

Then the bomb exploded.

* * *

I felt the shock wave, which battered everything in the room. It hit me like God's huge fist.

The floor shook. The whole building shuddered. I couldn't hear a thing, though. The explosion blew out my eardrums.

But it didn't kill me. Not yet.

I lay there on my back, in the center aisle of the auditorium, while bits of debris pelted my face. Terrified, I waited for the crashing to stop.

Then I opened my eyes.

All the lights had gone out. The auditorium was dark and smoky. The pain in my leg was still unspeakable, and blood was still pumping out of me. And something heavy sat on my chest, making it hard for me to breathe.

I needed to push the weight off me. But as soon as I raised my hands and touched the thing, I felt a rush of horror.

It was Ben. His body lay facedown on top of me, motionless, his torso perpendicular to mine. My right hand brushed his neck, and my left touched something firmer, a splintery object that I soon realized was a long slab of wood. It stuck out of Ben's back, almost vertical, its bottom end buried in the flesh between his shoulder blades. It was a piece of the lectern. The explosion had blown it across the room and impaled him.

I gagged. I turned my head to the side and let out a dry heave. Then I started coughing.

The smoke in the room was getting thicker. I peered into the darkness and saw flames at the front of the auditorium. The blast had ignited the stage curtains and floorboards, and the burning debris from the explosion was spreading the blaze. Several rows of wooden seats had already caught fire. I coughed harder, struggling for air.

I was trapped. I'd lost too much blood, and now I was too weak to crawl away. I barely had the strength to lift my head off the floor. And though the room was getting hotter and hotter, I kept on shivering. I was going into shock.

And I was alone. No friends. No army of supporters.

Where's God? Where's the Adversary? Where are the saints and angels, the holy knights and archers?

I was losing it. My mind was melting. The fire came closer, racing across the auditorium, the rows of seats erupting in flames, one after another.

Was any of it true? Did I really see the Lord? Or did I just go crazy, like all those other mathematicians?

The seats on both sides of the aisle were ablaze. I lay on the floor between two raging fires, and the smoke was like molten lead in my throat. I was going to burn to death, just like Joan of Arc. And just like my sister.

Samantha! Help me! SAVE ME!

My sister stood in the aisle, her blond ponytail swinging. She bent over and embraced me, pressing her chest against mine. She kissed the side of my head and whispered into my ear.

"Be of good cheer, Joan. I know you can do this."

HELP ME, SAMMY! GET ME OUT OF HERE!

She hugged me tighter. "I hate to give you more bad news, but there's no afterlife for humans."

No. I was hallucinating again. It wasn't Samantha.

Her voice and words came from my own memory, and the pressure against my chest was really the weight of the corpse, the body of the boy who'd tried to save me. Poor Ben, who was so cruel to me at first but then changed his mind and defended me. He turned away from evil and chose to be good, which is something that only human beings can do. Because we invented good and evil. God had nothing to do with it.

The fire arched over the aisle. The flames drew toward me from the left and right, scorching my arms, browning my shirt. I screamed again.

God! Do you hear me? You're nothing! You're worthless! Any human is a hundred times better than you, because WE CAN FEEL!

The smoke seared my lungs. My skin charred. My hair ignited.

I was a blazing, screaming creature of pain, flailing my blackened limbs in fury.

Then a savior appeared, a human savior, dressed in a firefighter's uniform. He grabbed my arms and pulled me out of the inferno.

Chapter Twenty-Nine

I spent the next nine hours in a hospital room. In a special unit for burn victims, I guess, probably somewhere in Albuquerque. I wasn't aware of my exact location, because I was *seriously* sedated.

I drifted in and out of consciousness. Everything came at me in bursts and flashes. A glimpse of a surgical mask. A pair of hands in latex gloves. The smell of ammonia, the glitter of scalpels. The pressure of bandages around my arms and legs and head. And the constant whoosh of air through the plastic tube that snaked into my mouth and down my throat.

On the plus side, though, I didn't feel any pain. They'd injected so much sedative into my bloodstream, I felt like I was floating in a warm, gluey pond. I probably would've floated right out of that hospital if not for all the tubes they'd stuck into me—the thick breathing tube that ran down to my lungs, the thinner tubes inserted into my nose and belly and crotch.

There was a ringing in my ears, but I could hear again. I could see too, through rectangular slits that had been cut into the bandages on my face. But I couldn't focus on anything for more than a few seconds. My mind kept drifting.

I saw Mom and Dad. Or at least I think I saw them. Dad bent over me, his face just inches from my eye slits. His cheeks were streaked with tears, but he was trying to smile. Mom sat beside my bed, one hand gripping the guardrail, the other resting on my bandaged arm. I felt pretty sure that I wasn't hallucinating them, because their presence in the hospital was perfectly logical. Charlotte probably gave the doctors my name and address, and my parents would've had enough time to fly to Albuquerque. But they looked older than I remembered, older and weaker. Mom's hand trembled as she patted my bandages. And Dad's face was so pale, his eyes so bloodshot. He looked like he belonged in his own hospital bed.

I saw Charlotte too. She stood on the left side of my bed, opposite from my parents, and pressed a wad of Kleenex to her beautiful eyes. Marguerite stood next to her, stiff and blank-faced, her sweatshirt gray with soot. She'd lost her brother, but she'd hated Ben so much that she must've found it hard to mourn for him. So she was here with me instead. She'd loved me for my mission, for what I symbolized: proof that God existed and that she was right to believe in Him.

But Charlotte's sorrow was deeper, because I was more than a symbol to her. We'd taken the first steps in an amazing journey. As I stared at her soot-streaked face and the crumpled ball of dirty Kleenex, all I wanted to do was touch her. I wanted to grasp her hand and squeeze it and tell her not to worry.

But I couldn't move. I couldn't talk either, not a single word. When I tried to say Charlotte's name, I gagged on the tube in my throat.

It was getting harder to breathe, even with all the oxygen that the ventilator machine was pumping into me. My lungs heaved and sloshed, their delicate lobes poisoned by the smoke, their airways clogged with dead tissue. After a while I realized why I had the sensation of floating in a warm pond. The dead tissue inside me was dissolving, and the thick fluid was filling my lungs. My mechanical breaths were getting shallower, because there was less and less room for air in my chest.

I was drowning. I was going to die.

That's why my parents looked so old and weak, so devastated. The doctors must've told them there was no hope. And now I knew it too.

If I weren't so groggy, I would've gotten furious. I'd had so many years ahead of me, so many dreams! I'd planned to go to Princeton and become a great mathematician. I was going to prove the Riemann Hypothesis and solve the Kobon triangle problem. And I would work in a beautiful office, like Professor Laura Taylor's at City College, with an awesome triangular desk and a framed portrait of my loving wife, whoever that might be. But none of those things was going to happen now, and I didn't even know who to blame. God? The Adversary? Vlad? Myself?

At that moment, though, I couldn't get angry. Or frightened. The sedatives in my blood took the edge off everything. I was just sad and confused. And very, very tired.

Then I saw Andrei.

He appeared at the foot of my bed, coming out of nowhere. One instant he wasn't there, and the next instant he was, dressed in his familiar polo shirt and khakis.

Neither my parents nor my friends noticed his sudden appearance in the hospital room. They didn't look up at him or say hello. They just kept staring at me and shaking their heads and crying.

Andrei bowed, very formally, and grinned his dopey grin. Then he spoke to me without moving his lips. The words didn't come out of his mouth or travel through the air. They went directly from his head to mine.

Hello, Joan. It's good to see you again. You're not feeling any pain now, are you?

It had to be a hallucination. No doubt about it.

No, I'm real. I changed the optics in this room to ensure that only you can see me. The others can't hear me either. I wanted to have a private conversation with you.

I still didn't believe it. But in my fragile condition—muzzled, choking, dying—I couldn't argue the point.

Actually, you *can* communicate your thoughts to me. Just compose the sentences in your head. You can still do that, yes?

This was absurd. I was too tired for his games.

Go away. Let me die in peace.

Andrei grinned again. He stepped closer to my bed and leaned over my bandaged feet. **There, you did it! So now we can talk. We have an important matter to discuss, Joan.**

You're not listening. I told you to go away.

Look, I understand why you're angry. I lied to you. I tricked you into participating in a dangerous plan. And it didn't go well.

He was disturbing my quiet descent into oblivion. Despite the fact that I had enough drugs in me to calm an elephant, my rage reawakened.

Didn't go well? Are you referring to the bomb? The explosion that turned me into a freakin' human torch?

I'm truly sorry about that. But remember, I tried to convince the Big Cheese to save you.

I was so angry, my brain sputtered. I could barely compose a sensible thought.

I don't believe this! You think you can shift the blame?

In this case, yes. The Master of the Universe really overdid it this time. If He wanted to protect His cosmic secrets, all He had to do was destroy your laptop and the math notebooks. But no, He let the stupid Russian pulverize the whole auditorium.

But you're the one who started the fight with Vlad in the first place! And tell me again why you put me in conflict with that sadist? Because you thought I needed some more motivation?

Andrei shook his head. **Believe me, if I were in control of things, you wouldn't have been hurt. I would've disarmed the bomb and jammed the Russian's gun. But I'm not the Almighty, you see? I'm only the Adversary. And when He and I confront each other, He usually wins.**

So what are you saying? You want me to feel sorry for you?

No. I want to make amends. I want to give you something that will make up for what you've lost.

Now I was confused again. My hallucination was torturing me.

You can't make amends. You killed me, okay? So unless you have a brand-new body for me, you should just get out of here.

I can't put you in a new human body. The Big Cheese would definitely notice something like that. But I'm keeping a secret from Him, a clever trick I've invented. I haven't tried it on anyone yet, but I'm pretty sure the procedure will work.

Please, stop talking. I've had enough of your tricks.

Do you remember when I said there were only two immortal beings in the universe? Well, I think I can change that. I can save your spirit, your unique intelligence. Under ordinary circumstances, it would vanish when you die, but I can take it out of your body before that happens. You'd become like me, incorporeal and immortal. Then there would be three of us.

Andrei leaned farther over my bed, looking straight at my eye slits. His face beamed with pride, the brightest thing by far in that hospital room.

I felt queasy. The warm fluid in my lungs rose higher.

I thought you said there was no afterlife for humans.

That's true. It's never happened before. You would be the first.

And you invented this...procedure? Like a recipe?

I've been working on it for a while, a few thousand years. Because I've never tested it, I can't guarantee that the process will be completely trouble-free. You might have some difficulty with the transition. But I have confidence in you, Joan. You're a very resilient person. In fact, you're the most remarkable human I've ever met.

He pointed at my bandaged body. He was flattering me again, trying to fool me.

I don't believe you.

He shrugged. **All right, I admit it, I'm bending the truth a bit. I've met several other remarkable humans over the centuries. There was one young man in particular whom I found very intriguing, a lively preacher from the Levant. I seriously considered making him immortal, but the Big Cheese killed him before I got the chance.**

So this is another scheme. You wouldn't really be doing it for my sake. You're looking for an ally in your war against God.

Andrei let out a sigh. **Call it whatever you want. The point is, I'm offering you immortality. You won't be as powerful as the Almighty, but He won't be able to eliminate you either. And you won't be under my control, not even for a second. You'll be free to do anything you please.**

But you're hoping I'll side with you, right? Because I'll feel an obligation? And because God is such a humorless, arrogant jerk?

He chuckled. **Yes, exactly. I'm also hoping that you'll make this universe more interesting. You're an entertaining person, very unpredictable. And I'm in desperate need of entertainment these days. Perhaps we can work together on a few projects? Maybe persuade the Big Cheese to change some of His plans?**

Grinning again, he spread his arms wide. It was the universal gesture of welcome, an offer to embrace me. All I had to do was say "Yes" in my head. Then the Adversary would lift my spirit from my broken body, and for the rest of eternity I could roam across the cosmos.

But would it really be that simple? Would the bargain be as free and easy as he claimed? Or would there be hidden terms and conditions? I had the feeling that Andrei hadn't told me everything. There were mysteries he hadn't explained, like why God and the Adversary were eternally at war, and why the Lord couldn't simply eradicate His enemy. And then there was the greatest mystery of all: who created the Creator?

My lungs gurgled. They were so full of fluid, the ventilator couldn't push any more oxygen into them. The machines in the hospital room blared their high-pitched warnings.

It's your choice, Joan, all yours. But you'll have to choose soon. Once your body dies, your spirit will expire too, and there won't be anything left for me to save.

I was suspicious. Because I knew it wasn't smart to make bargains with the devil. He was the Prince of Lies, after all. I couldn't trust him.

But immortality is a pretty attractive temptation. Especially when the only alternative is nothingness.

I weighed my options. The room darkened. I closed my eyes.

Epilogue

Dr. Peter Cauchon stared across his desk at Joan Cooper's father. The poor man sat in the same leather chair his daughter had occupied during her brief visit to the office three months ago.

"Thanks for fitting me into your schedule, doctor." Cooper leaned forward in the chair, clearly agitated. His hands were clenched, his knees jiggled. "I really appreciate it. I know you don't have many adult patients. You treat mostly teenagers, right? That's what I read on your website." His eyes darted back and forth, gazing for a moment at each of the book-lined walls. "But I think you're the best therapist for me right now. Because you met Joan. You talked with her. For a few minutes, at least."

Dr. Cauchon examined him. Cooper seemed tense, troubled. His hair had turned white since last October. He wore a wrinkled plaid shirt that was missing a button, and his face was gaunt and covered with gray stubble. He'd brought a backpack into the room, an old, black thing that slumped on the floor beside his scuffed shoes. He looked distressed and disheveled and very out of place in Cauchon's office. The doctor charged a premium rate for his services, and his patients—even the surliest teenagers—were usually clean and well dressed.

Cauchon winced. He'd made a mistake. If he'd known that Cooper was in such a precarious state, he would've never agreed to see him. The man obviously needed a more appropriate psychiatrist, one who specialized in grief counseling. But now it was too late to cancel the appointment, so the only option was to hear him out for the next forty-five minutes, then suggest the name of another therapist.

"All right, Mr. Cooper, I'm here to listen. Can I call you by your first name? It's Joe, isn't it?"

He nodded. "Yes, yes, that's right. I've thought about you a lot, doctor, and how you tried to help Joan. It wasn't your fault that it didn't work out. We should've brought her to you sooner. We just didn't know how…how bad she…"

The man unclenched his hands and turned the palms up, as if he were expecting something to fall into his lap. He sat there with his mouth open, but said nothing. After a few seconds the silence became uncomfortable.

Cauchon cleared his throat. "I realize how difficult this is, Joe. To lose a child is horrible enough. But to lose two daughters, one right after the other? The pain must be unimaginable." He tilted his head and tried to look sympathetic. The feeling didn't come to him naturally, but with a little effort he could mimic it. "How are you and your wife handling the situation? Is it getting any easier as the days go by?"

There was another silence, even more awkward this time. Cooper bit his lip. "Mary and I…we separated last month. We weren't helping each other anymore."

"I'm so sorry that—"

"Every time we talked about…what happened to Joan…we just made it worse. We argued. We blamed each other. It wasn't good."

His voice cracked. It sounded like he was about to start crying. Cauchon maintained his sympathetic expression, but underneath it he felt a twinge of annoyance. He could tolerate the sight of teenagers crying, but he had no patience for weeping adults. He needed to head off the outburst. "Well, the important thing is that you're seeking help. That's a good sign, a very healthy sign. It means you're confronting the challenge."

Cooper shook his head. His eyes were glassy, but fortunately they didn't well with tears. "No, it's more complicated than that. I think there might be something wrong with me. With my mind."

Cauchon raised his right hand from the desk and cupped his chin. It was a contemplative gesture, intended to convey that he was thinking deeply about what his patient had just said. "I understand, Joe. Sometimes grief can get so overwhelming, it knocks you right off your feet. But the essential *you* is still there, under all the emotion." He leaned across the desk. "You've experienced one of life's most agonizing hardships. That can make anyone feel lost and wrong."

Cooper shook his head again. At the same time, he bent over and reached for the backpack on the floor. "I want to show you something." He unzipped the bag and pulled out a laptop. Then he stood up and came around to Cauchon's side of the desk. "This file appeared in my Documents folder three days ago."

He placed the laptop on the desk in front of Cauchon and opened it. But the doctor turned away from the screen. Although he recognized that Cooper was distraught, he hated it when patients invaded his personal space like this. "Joe, please sit down. We can—"

"There! See it?" He pointed at the screen, his finger almost touching the glass.

"Listen, I realize you're upset, but this isn't—"

"Just look at it! Look at the name and date of this file."

Very reluctantly, Cauchon turned to the screen. There was a long list of file names inside the window of the laptop's Documents folder. Halfway down the list, one of the names was highlighted in blue: **Joan137**. Cauchon looked at the "Date Modified" section next to the name and saw that the file was last changed on October 25 at 6:32 pm.

Cooper edged closer to him, still pointing at the screen. "That's the day she died. *The exact moment.* But I didn't see this file on my computer until last Sunday. It just popped into existence, out of nowhere. It wasn't there when I looked at my Documents folder on Saturday, but the next day I saw it. And I still have no idea where it came from." He bent over the laptop's keyboard and pressed the touchpad. "Now I'll open the file. Look at this."

After a two-second wait, a Word document appeared on the screen. Cauchon read the first paragraph.

I'll start at the beginning, okay? My name is Joan Cooper, and I'm seventeen years old. I'm going to tell you about the first time I saw God.

He skimmed the following paragraphs. It seemed to be a journal written by the unfortunate girl. The file was long, hundreds of pages.

Cooper bent lower. His shoulder brushed against Cauchon's, but his eyes stayed on the screen. "This is Joan's story, doctor. She wrote it for me. And somehow she managed to get it on my computer."

Cauchon grimaced. Cooper's voice was too loud. Up close, he looked feverish—his forehead glistened with sweat, and his jaw muscles quivered. Cauchon leaned to the right, trying to put a few inches of space between them. "Have you, uh, read this whole document?"

He nodded. "It starts a week before she died. And it describes everything. All her visions, all her discoveries."

"When did she write it? Before she drove off to New Mexico with her friends?"

Cooper turned his head and stared at him. His eyes shone in the bluish light from the screen. "That's the strange thing. The thing that's driving me crazy." He pointed again at the computer. "The story goes right to the very end. She describes…" His voice cracked again. "She describes her own death."

Cauchon leaned farther away. The odor of the man's sweat was overpowering. "Please, Joe, if you could just go back to the chair and—"

"She couldn't have written it in advance. She tells everything *exactly* as it happened. How she marched to the convention center in Albuquerque. How she warned everyone to leave the auditorium." Cooper raised both his hands to his head. He pressed his palms against his forehead and laced his fingers into his white hair. "And how she lay in the hospital bed afterward, in the burn unit. I mean, I was there, I saw it. How she struggled to breathe. How she…"

His voice trailed off and his face reddened. Cauchon was concerned now, and a little frightened too. Cooper seemed to be suffering from emotional dysregulation. There was a good chance he might act impulsively and lash out at the nearest target. It was imperative to calm him down.

"Joe, please relax. I'm sure there's a simple explanation for all this."

"The explanation is in the last chapter of her story. But it's not simple at all." He slid his hands upward, clamping his head on both sides. His fingers dug into his scalp. "Joan was resurrected. Not by God, though. By the Adversary."

Cauchon chose to ignore this. He had no intention of starting a long-term relationship with this patient, so there was no point in delving into the man's nonsense. He waved it aside. "I'm thinking of another explanation, something more reasonable. For instance, someone could've hacked your computer. Any malicious hacker who knows your email address could've slipped that document into your laptop. That would explain everything."

Cooper narrowed his eyes. His hands were still twined in his hair. "A hacker?"

"Yes, it's a possibility. In her last few days, Joan was associating with some odd characters, wasn't she? Religious zealots, New Agers? That's what all the newspapers said." Cauchon thought back to the news stories he'd seen after the bombing in Albuquerque. He was going to use plain, irrefutable facts to dispel Cooper's hysteria. "One of those zealots could've written the story about Joan and then sent it to you. Maybe out of a misguided religious impulse. Or maybe it's just a prank. Those fanatics can be very cruel, you know."

Cooper went silent for several seconds, his lips trembling. Then he shook his head. "No. Joan wrote the story. It sounds just like her." He lowered his hands and turned back to the laptop. "She's in heaven. The Adversary offered to resurrect her spirit, and she agreed. Then she sent me the story. To prove that she survived."

"I don't think—

"Now she's the third immortal being in the universe. And she wanted me to know it, so she put her story on my computer. Because she could see how unhappy I was, and she wanted to help me. It's exactly the kind of thing that Joan would do."

Cauchon took a deep breath. He couldn't ignore the problem any longer. Cooper seemed to be suffering from psychosis as well. This was going to be tricky. "Joe, I have to ask you a question. I don't mean to upset you or cast any doubt on your honesty. I just want to get to the bottom of this matter." He tilted his head toward the laptop's screen. "Did *you* write this story? Maybe as a way to keep Joan's memory alive?"

The man fell silent again. To Cauchon's relief, he took a step backward. "I don't remember writing it."

"Are you sure? Think it over, Joe. Perhaps you're blocking the memory. That happens sometimes, especially in cases of extreme distress."

Cooper took another step backward. He swayed on his feet. "So that's your diagnosis? You think I'm going nuts?"

"Now, now, I never said—"

"If I'm doing things without remembering them, isn't that the very definition of crazy?"

Cauchon frowned. He was getting tired of this. He wished he'd never heard of the Cooper family. "Look, I can't make a diagnosis at this stage. And to be honest, I don't think I'm the right therapist for you either." He reached for the iPad on his desk. It held a Word file containing the contact information for every psychiatrist in New York City. "You need a grief counselor. Someone who can help you process the loss of your daughters. Can I recommend an excellent professional for you?"

Cooper didn't respond at first. He just stared straight ahead and backed up against the bookshelves behind him. He looked unsteady, and for a moment Cauchon expected him to crumple to the floor. The doctor readied himself to call his receptionist for help.

But instead of collapsing, Cooper came back to the desk. He grabbed his laptop, closed it, and tucked it under his arm. Then he turned around and marched out of the room. He left Dr. Cauchon's office without saying another word.

Just like his daughter.

* * *

Two minutes later, Joe stood at the corner of 82nd Street and Central Park West, breathing hard and leaning against a lamppost. The people on the sidewalk stared at him as they walked past, but no one stopped. He supposed they were afraid to approach him. He was crazy, after all. And in New York, everyone steers around the crazy people.

No. I'm not crazy. I know I didn't write that story, because I couldn't have written it. It kills me to think of Joan for even a minute, even a second. So how could I have written all those hundreds of pages about her?

He tightened his grip on the lamppost, which supported a traffic signal as well as a streetlight. It was a cold morning in January, and the frigid gray steel numbed Joe's fingers, but he held on to it anyway and stared at the cars on Central Park West. The taxis sped through the filthy slush left over from last night's snowfall. An M10 bus hurtled toward the stop on 81st Street.

Okay. Okay. I didn't write the story. And it didn't come from any religious zealot hacker either. But does that mean I should believe all this stuff about God and the Adversary? And their conflicting plans for the universe? The story is completely nuts. It's crazier than anything in the Bible.

If only Joan were here. She would tell him what to do. Ever since she was eight years old, she'd given him good advice. He hadn't realized how much he relied on her until she was gone.

Joan? Are you in the ether somewhere? In the eleventh dimension, the black void beyond the cosmos? Are you plotting rebellion with the Adversary and battling against God? Or is your story just a fairy tale, just as absurd as all the other hopeful myths?

Joe waited for an answer. He tuned out the noise of the Manhattan traffic and listened for his daughter's voice.

It's a yes-or-no question, Joan. Should I believe?

There was no answer. Aside from the cars and buses, he heard only the drone of a distant airplane. And, a moment later, the whistle of a sparrow in Central Park.

But he didn't need to hear Joan's voice. He realized that he already knew the answer. It was encoded in her story like a secret message, like a mathematical formula. The same formula that governs the movements of the stars and the oscillations of every atom and molecule.

Joe decoded the message. His pulse raced. It was so clear to him now, so obvious.

All our assumptions about God and the universe are wrong. Our great task is to understand the cosmos, and our best tool is mathematics.

Joan's string-theory equations were lost in the Albuquerque bombing, but it was indisputable that she'd found the correct solution to the problem. The proof was that she'd calculated the precise value of alpha. So there was a chance that physicists could reconstruct her solution. And maybe she'd planted a few important clues in the outlandish story she'd slipped into Joe's laptop.

Joe let go of the lamppost and steadied himself on the street corner. He could get in touch with the string theorists who'd come to Joan's presentation in New Mexico. He could show them her story and see what they thought of it. Although she hadn't left behind any sketches of her Calabi-Yau manifold, she'd devoted several paragraphs to describing its shape. Maybe that would be enough to inspire the physicists. Maybe it would give them new hope and a new direction.

But Joe shivered as he considered the plan. Was it really a good idea to pursue cosmic understanding? Would the benefits of the new knowledge outweigh the hazards? Joe had read his daughter's story very carefully, paying special attention to the warnings of catastrophe and the promises of peace, yet he still didn't know what to do. Which was more dangerous, ignorance or enlightenment?

Joe stood there, unmoving, for almost a minute. He stared at the park across the street, the bare trees stretching their limbs, the park benches heaped with snow. The decision was too much for him. He wished that someone else would step in.

Then, during a lull in the traffic noise, he thought he heard something.

It's your choice, Dad.

All yours.

Author's Note

Just before I started writing this novel, I had a conversation with a Catholic friar on the Upper West Side of Manhattan. I spotted him one afternoon on Amsterdam Avenue: a middle-aged man wearing glasses, a baseball cap, and a gray, ankle-length religious habit, bound at the waist with corded rope.

It was a pretty unusual sight for the neighborhood, which has its fair share of churches and synagogues but isn't particularly religious. What made it even more striking, for me at least, was that I'd just spent several days reading about the history of the Catholic Church. My planned novel was going to tell the story of a modern-day Joan of Arc, and I'd fallen into the trap of doing way too much research for the book. I'd gotten sidetracked by the stories of the medieval saints, especially St. Thomas Aquinas and St. Francis of Assisi. And now here on the street was an embodiment of the subject I was exploring, an honest-to-goodness Franciscan friar on a New York City sidewalk, strolling past a pizzeria and an eyebrow threading salon.

I couldn't help but stare. Luckily, he didn't take offense. He just smiled and said, "Hello! Would you like to ask me a question?" My answer was yes. I had a million questions for him.

After a few minutes, I told him about my idea for a novel. I thought the friar might be horrified, but instead he was encouraging. He suggested that I read Mark Twain's *Personal Recollections of Joan of Arc*. He also recommended that I see the Broadway revival of George Bernard Shaw's play *Saint Joan*, which is especially good at dramatizing Joan's trial and execution. (The fine actress Condola Rashad starred in that production, while Walter Bobbie played the role of the traitorous French bishop Pierre Cauchon.)

M. Alpert, *Saint Joan of New York*, Science and Fiction,
https://doi.org/10.1007/978-3-030-32553-4_3

Last but not least, the friar recommended *A Still, Small Voice* by Benedict J. Groeschel, one of the founders of the New York-based Community of the Franciscan Friars of the Renewal. This book provides practical guidance on how to evaluate the claims of people who say they've received divine revelations. As I wrote my novel about a modern-day Joan of Arc, I began to see similarities between Joan Cooper's strange visions and those described in Groeschel's book. In my novel, Joan gets a glimpse of the extra dimensions hypothesized by string theory, which posits that the shape of those dimensions—they can be folded into a huge variety of microscopic Calabi-Yau manifolds—determines all the laws of physics in our universe. If this all-important hidden geometry appeared to a mathematician or physicist in a vision, they might very well conclude that the revelation had come from God.

Before we wrapped up our conversation on Amsterdam Avenue, the friar offered to say a prayer for me. I said yes again; although I'm not Catholic, how could I turn down such a nice gesture? It was a great prayer, really heartfelt, and at the end of it the friar invoked Saint Joan, asking her to bless my project.

If Joan does still exist somewhere—who knows?—I hope I haven't let her down. As I worked on my novel, I sometimes imagined her looking over my shoulder and scowling at all the heretical things I was writing. But I also imagined that she would be patient with me. Surely, someone who was put to death for her intransigence, her brave refusal to back down from her idiosyncratic beliefs, would forgive a writer for envisioning a metaphysics that's a bit different from the usual depictions.

That's my hope at least.